T0074962

PROPAGATION OF
PACIFIC NORTHWEST NATIVE PLANTS

Publication of this book was made possible
in part by a grant from

USDA Forest Service
Cooperative Programs
Pacific Northwest Region

PROPAGATION OF
PACIFIC NORTHWEST
NATIVE PLANTS

Robin Rose
Caryn E.C. Chachulski
Diane L. Haase

Oregon State University Press
Corvallis, Oregon

The paper in this book meets the guidelines for permanence and durability of the Committee on Production Guidelines for Book Longevity of the Council on Library Resources and the minimum requirements of the American National Standard for Permanence of Paper for Printed Library Materials Z39.48-1984.

Library of Congress Cataloging-in-Publication Data
Rose, Robin, Dr.
Propagation of Pacific Northwest native plants / Robin Rose, Caryn E.C. Chachulski, and Diane L. Haase.
 p. cm.
Includes bibliographical references and index.
ISBN 0-87071-428-7 (alk. paper)
1. Native plants for cultivation—Northwest, Pacific.
2. Native plants for cultivation—Propagation—Northwest, Pacific.
I. Chachulski, Caryn E.C. II. Haase, Diane L. III. Title.
SB439.24.N673R67   1998
635.9'51795—dc21   97-41008
                                                              CIP

© 1998 Robin Rose, Caryn E.C. Chachulski,
and Diane L. Haase
All rights reserved.
First edition 1998, fifth printing 2005
Printed in the United States of America

Oregon State University Press
102 Adams Hall
Corvallis OR 97331-2005
541-737-3166 •fax 541-737-3170
http://oregonstate.edu/dept/press

Table of Contents

Introduction

Suggested Readings

Bluhm, W. 1992. Basic principles for establishing native grasses. *Hortus Northwest* 3:1-3.

Carlson, J.R. 1992. Selection, production, and use of riparian plant materials for the western United States. pp. 55-67 *In*: Landis, T.D. (tech. coord.) Proceedings, Intermountain Forest Nursery Association. Park City, UT; August 12-16, 1991. Ft. Collins, CO: USDA Forest Service, Rocky Mountain Forest and Range Experiment Station. Gen. Tech. Rep. RM-211.

Dirr, M.A., and C.W. Heuser. 1987. *The Reference Manual of Woody Plant Propagation: From Seed to Tissue Culture.* Athens, GA: Varsity Press, Inc. 239p.

Evans, J.M. 1992. Propagation of riparian species in southern California. pp. 87-90 *In*: Proceedings, Intermountain Forest Nursery Association. Park City, UT; August 12-16, 1991. USDA Forest Service, Gen. Tech. Rep. RM-211. 119p.

Native plants have been increasingly recognized as a crucial component of forest management, especially after a fire or some other disturbance. They provide many benefits to the forest ecosystem such as erosion and flood control, wildlife forage and habitat, species diversity, soil stabilization, aesthetic enhancement, riparian restoration, revegetation of road cuts, and improvement of recreational areas.

In order to successfully propagate native plants, it is important to know a good deal about the target species. For seed collection, the plant must be properly identified and seed harvest must occur at the appropriate time for optimal seed vitality. The time of harvest can vary over the geographical range for a single species. Some species only produce an adequate seed crop every few years while others have prolific seed production every year. The method of seed collection can vary greatly among species based on plant form and seed size.

Growing native plants from seed can be a complicated task. Once collected, the variety of fruits, acorns, pods, capsules, and berries require differing equipment and techniques for extraction. Furthermore, seed longevity varies greatly among species: some can be stored for years while others need to be germinated immediately. The success of germination depends on knowing what each species requires to overcome physical barriers (i.e., seed coat) or physiological barriers (i.e., dormancy). The appropriate media, moisture, temperature, scarification, chemicals, light, and nutrients determine the success of germination and subsequent growth.

Vegetative propagation of native plants can also pose some interesting challenges. Plants can be produced by cuttings, division, layering, rhizomes, tissue culture, and grafting. Again, an understanding of each species is important to successful propagation by these techniques. One of the most common techniques is by rooted cuttings, but this must be approached on a species by species basis. Some species root more successfully using branch tips while others root well with stem, leaf, or root cuttings. In addition, many species root more readily when treated with a root growth hormone.

The proper culturing of native plants, whether propagated by seed or vegetatively and whether grown in containers or bareroot, is another essential step to ensuring a vigorous plant crop. A certain level of heat and humidity is often required during germination of seed or rooting of cuttings. In addition, a careful fertilization and irrigation regime must be followed for good plant development and proper phenology. It is important to know which plants can withstand stress and which are extremely sensitive to variations in moisture, temperature, light, or other environmental factors.

Obviously, native plant propagation requires some experimentation and innovation. With so many species-specific propagation requirements and very little information available in the literature, native plant growers must refine their techniques based on trial and error and their available equipment, supplies, and facilities. Furthermore, the final product must be based on the ultimate use of the plant. Very large root systems may be desirable for planting in sand banks while a tall shoot may be needed to compete with surrounding vegetation in a riparian environment. In other cases, a small seedling may be sufficient to meet outplanting goals or perhaps seedlings from more than one age or size class may be used. To achieve the desired plant specifications, a plant grower must allow for the necessary nursery space and growing period.

Until now, there has not been a comprehensive manual on propagation of Pacific Northwest native plants. Until recently, most articles about native species have suggested ways to control or eradicate them. Increasing awareness of their beneficial role in promoting a healthy, stable ecosystem has necessitated a more detailed and extensive information base for their propagation. A thorough search of forestry and agricultural journals as well as gardening and horticultural handbooks does yield some useful propagation information, but such an exhaustive literature search is not practical or convenient for many who wish to grow native plants. Furthermore, some of

Hortus West: A western North America native plant directory and journal. Wilsonville, OR: Pacific Habitat Services Inc., Communication Services Division.

Lohmiller, R.G., and W.C. Young. 1972. Propagation of shrubs in the nursery. pp. 349-58 *In*: Wildland shrubs—their biology and utilization: an international symposium. Utah State University, Logan, UT; July 1971. USDA Gen. Tech. Rep. INT-1.

Macdonald, B. 1986. *Practical Woody Plant Propagation for Nursery Growers.* Vol. 1. Portland, OR: Timber Press. 669p.

Mirov, N.T., and C.J. Kraebel. 1939. Collecting and handling seeds of wild plants. U.S. Civilian Conservation Corps. Forestry Publication No. 5. 42p.

the best information is in the minds of those who have learned through direct experience, many of whom have not had the time, funding, or inclination to publish or have only done so on a very limited basis.

The purpose of this manual is to present a compilation of information from literature sources and personal contacts and to make it widely available. However, because this is a newly documented field, some of the information is still fairly general. We would like to make clear that the techniques we present in this manual are not necessarily the only successful ones for a given species. We encourage readers to experiment with propagation techniques. If you discover a successful technique, tell us and we'll include it in the next edition of this manual!

Metric Conversions

When you have	multiply by	to find
centimeters (cm)	0.3937	inches (in)
decimeter (dm) = 10 cm or 0.1 m		
millimeter (mm) = 0.1 cm		
meters (m)	3.2808	feet (ft)
kilograms (kg)	2.2046	pounds (lb)
seeds/kg	0.4536	seeds/lb
kg/ha	0.8929	lb/acre
seeds/m^2	0.093	seeds/ft^2

When you have	multiply by	then add	to find
temperature (°C)	1.8	32	temperature (°F)

For example:

30°C	=	86°F
20°C	=	68°F
10°C	=	50°F
0°C	=	32°F
-10°C	=	14°F

General Propagation Techniques
Tom Landis and Diane Haase

Plant propagation is both a science and an art. Specific technical skills must be acquired through innate ability or experience, and often require a certain "feel." Good plant propagators are said to have a "green thumb." Successful propagators must not only be able to reproduce the desired plant species but to do so economically and consistently.

The first phase of the planning process is to determine which propagation method will be most effective and economical for the plant species. Both the biology of the species and the objectives of the outplanting project must be considered. Growers can gather information from the published literature, compare propagation methods for related species, study the native environment and natural growth habit of the plant, and consult with other people growing the same or similar species.

In nature, some plants spread by vegetative propagation but propagation by seed is much more common. The choice of propagation method depends on the management objectives and the characteristics of the plant species. If the objective is to generate a large number of plants that have been genetically selected for some characteristics like fast growth or resistance to insect or disease, then vegetative propagation is a logical choice. On the other hand, if the objective is to produce plants that maintain the broad range of genetic diversity that is found in nature, then seed propagation makes more sense.

Seed Propagation

Seed propagation is encouraged whenever possible because it is easier to capture and preserve genetic variation with this method than with vegetative propagation. In addition, seed propagation preserves the wide genetic adaptation that is critical to successful seedling establishment and growth in the natural environment. The primary objective of seed propagation is to promote rapid and complete germination and establishment in the growth container.

The seeds of different plant species vary considerably in size, appearance, and internal anatomy but each consists of three basic parts: embryo, food storage tissues, and seed coat. The embryo is the miniature new plant which will eventually develop into the seedling. The embryos of some species are undeveloped, whereas others look like miniature seedlings with seed leaves

References and Suggested Readings

Beagle, G., and J. Justin. 1993. Using a constructed wetland to treat waste water and propagate wetland species. *Tree Planters' Notes* 44(3): 93-97.

Belcher, E. 1985. Handbook on Seeds of Browse-Shrubs and Forbs. Atlanta, GA: USDA Forest Service, Southern Region. Technical Publication R8-TP8. 246p.

Bonner, F.T. 1990. Storage of seeds: potential and limitations for germplasm conservation. *Forest Ecology and Management* 35:35-43.

Bonner, F.T., J.A. Vozzo, W.W. Elam, and S.B. Land, Jr. 1994. Tree Seed Technology Training Course. USDA Forest Service, Southern Forest Experiment Station. GTR-SO-106.

Carlson, J.R. 1992. Selection, production, and use of riparian plant materials for the western United States. pp. 55-67 *In*: Landis, T.D. (tech. coord.) Proceedings, Intermountain Forest Nursery Association. Park City, UT; August 12-16, 1991. Ft. Collins, CO: USDA Forest Service, Rocky Mountain Forest and Range Experiment Station. Gen. Tech. Rep. RM-211.

Date, L. 1994. Propagation of Pacific Northwest native plants from seed. Combined Proceedings, International Plant Propagators' Society 43: 299-300.

Dirr, M.A., and C.W. Heuser, Jr. 1987. *The Reference Manual of Woody Plant Propagation: From Seed to Tissue Culture.* Athens, GA: Varsity Press, Inc. 239p.

Dumroese, R.K., D.L. Wenny, and K.E. Quick. 1990. Reducing pesticide use without reducing yield. *Tree Planters' Notes* 41(4): 28-32.

(cotyledons) and an undifferentiated stem-root axis consisting of the hypocotyl and the radicle (see glossary for definition of these and other terms). Food storage tissues provide a source of nutrition for seed germination and seedling establishment. The seed coat provides mechanical protection for the embryo and prevents desiccation.

Seed collection

Seed source—the geographical area in which the seed was collected—is of paramount importance when propagating native plant species from seed. Seed source affects seedling performance in two ways: cold tolerance and growth rate. In general, seedlings grown from seed collected from higher latitudes or elevations will grow slower but tend to be more cold hardy than those grown from seed from lower elevations or more southern latitudes. So, it is usually best to collect seed from the same area in which the plants are to be outplanted.

The most comprehensive information on woody plant seed collection and processing can be found in *Seeds of Woody Plants in the United States* (Schopmeyer 1974) or *Seeds of Woody Plants in North America* (Young and Young 1992).

Seed cleaning

Nature has designed fruits and seeds to assure the wide dissemination and successful establishment of seedlings, but many of these ecologically useful adaptations are actually a hindrance to easy seed propagation by humans. For propagation purposes, the fruit serves no useful function and some fruits actually serve the ecologically useful, but horticulturally frustrating, function of delaying seed germination. Viable seeds need to be separated from their fruits and cleaned.

The basic seed-cleaning machine is the air-screen cleaner, which uses a combination of screening and air flow to remove debris and can be used to upgrade seedlots by three physical properties: size, shape, and density (Bonner et al. 1994). Another quick and simple way to remove empty seeds is by flotation. In general, heavier filled seeds will sink in water and lighter empty or damaged seeds will float. However, some filled seeds will float if water bubbles are trapped on the seed coat and empty seeds may sink if they are dirty. This procedure can be checked by cutting a sample of seeds open and determining the percentage of filled seeds.

Seed storage

Successful storage requires knowledge of the seed characteristics of different species as well as of the quality of the seed before storage, its moisture content, and the storage temperature and method. Thick or hard seed coats restrict moisture loss and seeds containing oils tend to be harder to store than those containing more starch. Seeds collected before they are mature or those that were under environmental stress during maturation do not store well. Seed handling before storage will also affect storage potential. Exposing seeds to direct sunlight or high temperatures, especially if they have high moisture contents, is damaging. The care with which fruits and seeds are handled and stored during collection, shipping, and processing is important. Careless processing can cause cracks in the seed coat or even bruise sensitive seed tissues. A cracked seed coat allows moisture to escape and provides an entry for fungal pathogens.

Moisture content is by far the most important factor in seed storage. Some seeds tolerate desiccation and therefore store easily for a relatively long time. Others are intolerant of desiccation and must therefore be stored above freezing. Temperature is the other important consideration. In general, the lower the temperature, the slower the rate of deterioration, although the recommended range differs by species. Most seeds can be stored at temperatures around freezing or slightly below for a short time. Storage containers must be rigid enough to provide physical protection during storage and handling. Although rectangular containers are more space efficient, round ones assure spaces needed for air circulation in the freezer. It is best to line cardboard or fiberboard containers with plastic bags.

Presowing treatments for nondormant seeds

Even seeds that are not dormant need to be treated before they can be sown. Seeds that have just been processed or have been in storage have lower moisture contents than is ideal for germination. Imbibition is the physiological process whereby seeds absorb the water necessary to start the metabolic reactions that lead to germination. This can be achieved by soaking seeds in water for 24 to 48 hours. A running water rinse is recommended to keep the dissolved oxygen content of the water high and avoid stagnated conditions. This treatment also softens the seed coat and cleans it,

Edson, J.L., D.L. Wenny, A. Leege-Brusven, R.L. Everett, and D.M. Henderson. 1994. Conserving threatened rare plants: some nursery strategies. pp. 150-57 *In*: Landis, T.D.; Dumroese, R.K. (tech. coords.) National Proceedings: Forest and Conservation Nursery Associations. Fort Collins, CO: USDA Forest Service, Rocky Mountain Forest and Range Experiment Station. Gen. Tech. Rep. RM-GTR-257.

Emery, D.E. 1988. Seed Propagation of Native California Plants. Santa Barbara, CA: Santa Barbara Botanic Garden. 115p.

Finnerty, T.L., and K.M. Hutton. 1993. Woody shrub propagation: a comprehensive approach. pp. 82-91 *In*: Landis, T.D. (tech. coord.) Proceedings, Western Forest Nursery Association. Fallen Leaf Lake, CA: September 14-18, 1992. Fort Collins, CO: USDA Forest Service, Rocky Mountain Forest and Range Experiment Station. Gen. Tech. Rep. RM-221.

Handreck, K.A., and N.D. Black. 1994. *Growing Media for Ornamental Plants and Turf.* Randwick, NSW, Australia: University of New South Wales. 448p.

Hartmann, H.T., D.E. Kester, and F.T. Davies, Jr. 1990. *Plant Propagation: Principles and Practices.* 5th Edition. Englewood Cliffs, NJ: Prentice Hall Career and Technology. 647p.

Hortus West: A western North America native plant directory and journal. Wilsonville, OR: Pacific Habitat Services, Inc., Communication Services Division.

Landis, T.D., and E.J. Simonich. 1984. Producing native plants as container seedlings. *In*: Murphy, P. (comp.) 1984. The challenge of producing native plants for the Intermountain area; Proceedings: Intermountain Nurseryman's Association conference. Las Vegas, NV: August 8-11, 1983. Ogden, UT: USDA Forest Service, Intermountain Forest and Range Experiment Station. Gen. Tech. Rep. INT-168. 96p.

removing possible chemical inhibitors or pathogens. Place seed loosely in a nylon mesh bag and then in a tank of water, adding enough water with a hose that the extra runs over the side.

Presowing treatments to overcome seed coat dormancy

Dormancy is an ecological adaptation that insures that seeds will only germinate when weather conditions, especially moisture and temperature, are favorable to the survival of the seedling. Seed coat dormancy is often called "external dormancy," because the restricting factor is the tissue surrounding the embryo. The degree of seed coat hardness varies among species, but also depends on the ecotype and on weather conditions while the seed ripens (Macdonald 1986). Species that are adapted to fire-dominated ecosystems, such as many of the chaparral species, have seeds that need hot water or acid scarification in order to break seed coat dormancy and allow germination in an artificial environment. Several treatments can be used to soften the seed coat but remember that the objective is to just increase the permeability to water and gases; overly severe treatments may injure the embryo.

Hot water soak. Soaking in hot water is the traditional treatment for many legume seeds and those with waxy seed coats. Bring a container of water to a boil, immerse the seeds, and then remove the container from the heat and allow it to cool. The water should be heated to only 65 to 70°C for those seeds whose embryo can be damaged by high temperatures. Remove and dry the seeds when they swell and become gelatinous to the touch. It is best to experiment with each species and seedlot because of the variation in the thickness of the seed coat. The treated seed is subject to bacterial and fungal infection and so should be sown within a few days.

One problem with seeds that have been treated with hot water is that they stick together. One remedy for this is to place the seeds in moist peat moss for a few days (Macdonald 1986).

Scarification. Weakening the hard seed coat just enough to allow imbibition is necessary to overcome seed coat dormancy in some species, and several techniques are effective. Small quantities of relatively large seeds can be treated by hand: nick them with a triangular file or sharp knife, rub them against coarse sandpaper, or burn them with an electric soldering iron or wood-burning tool. To treat large seedlots, a rotating drum lined with sandpaper or a cement mixer filled with gravel can be used. Macerating berries in a modified household blender can be an effective mechanical scarification technique. Dull the blades with a file and blend the fruit in two to five volumes of water. Whatever technique is used, it is important to regularly check the seed coats to make sure that the treatment has not gone too far.

Another scarification method is to soak the seeds in a strong acid solution. Concentrated sulfuric acid is preferred but growers must be aware that this is an extremely caustic material and that safety always must be a paramount consideration. When properly done, acid scarification is a very effective way to remove hard seed coats and stimulate quick germination. Seeds should be clean and dry. Place them in the container, pour the acid slowly over them, and let them soak for 15 to 90 minutes. Because the treatment time will vary considerably with species and seedlot, it is a good idea to conduct some small-scale trials first by removing a few seeds at regular time intervals and cutting them to assess the thickness of the seed coat. Macdonald (1986) presents a good operational procedure for acid scarification.

Presowing treatments to overcome embryo dormancy

The cultural treatment for "internal" types of dormancy must overcome a physiological or morphological condition within the seed itself. The degree of dormancy can vary considerably from species to species among ecotypes, and so, again, the need to try different treatments and keep good records cannot be overstressed. Most species from the temperate zone require some exposure to the cold temperatures and moist conditions that occur naturally during winter; their seeds therefore require a cold, moist stratification treatment before the seeds will germinate.

Stratifying seed under cold and moist conditions is the most common treatment to overcome seed dormancy. Cold, moist stratification satisfies several important physiological functions, including activating enzyme systems and converting starches to sugars for quick metabolism. Even species that do not exhibit true

Macdonald, B. 1986. *Practical Woody Plant Propagation for Nursery Growers.* Vol. 1. Portland, OR: Timber Press. 669p.

McClelland, M.T., and M.A.L. Smith. 1994. Alternative methods for sterilization and cutting disinfestation. Combined Proceedings, International Plant Propagators' Society 43: 526-30.

Mountz, R.D. 1993. Greenhouse and Shadehouse Production Manual. Mason, IL: Illinois Dept. of Conservation, Mason State Nursery. 34p.

Munson, R.H., and R.G. Nicholson. 1994. A germination protocol for small seed lots. *Journal of Environmental Horticulture* 12: 223-26.

Schopmeyer, C.S. (tech. coord.) 1974. *Seeds of the Woody Plants in the United States.* Agric. Handbook 450. Washington, DC: USDA Forest Service. 883p.

Stein, W.I, R. Danielson, N. Shaw, S. Wolff, and D. Gerdes. 1986. Users Guide for Seeds of Western Trees and Shrubs. Corvallis, OR: USDA Forest Service, Pacific Northwest Research Station. Gen. Tech. Rep. PNW-193. 45p.

Tinus, R.W., and S.E. McDonald. 1979. How to Grow Tree Seedlings in Containers. Fort Collins, CO: USDA Forest Service, Rocky Mountain Forest and Range Experiment Station. Gen. Tech. Rep. RM-60. 256p.

Wasser, C.H. 1982. Ecology and Culture of Selected Species Useful in Revegetating Disturbed Lands in the West. USDI Fish and Wildlife Service, Pub. No. FWS/OBS 82/56. Washington, DC: U.S. Government Printing Office. 347p.

Young, J.A., and C.G. Young. 1992. *Seeds of Woody Plants in North America*. Portland, OR: Dioscorides Press. 407p.

dormancy may have faster and more complete germination from cold, moist stratification. Another cultural benefit is that early germination at low temperatures is delayed, allowing an even flush of germination when the sown containers are placed in a warm propagation environment (Bonner et al. 1994).

Some nurseries mix seeds with damp peat moss in a plastic bag which is then placed in a refrigerator. The condition of the seeds is checked weekly, and they are sown after the prescribed stratification period, or planted as germinants (pregerminated seeds). Another method is naked stratification: soak seed in water to obtain full imbibition, drain off the excess water, and place the seeds in polyethylene bags in refrigerated storage where the temperature is held slightly above freezing. Running water rinses are preferred to standing soaks because the bubbling water keeps dissolved oxygen levels high and cleanses the seed coat of pathogenic organisms. The best temperature for cold, moist stratification depends on the species and ecotype. Most trees and shrubs from colder climates need temperatures slightly above freezing; the optimum temperature range for most temperate-zone species is 1 to 5°C. The length of the cold stratification treatment can vary from four to twenty weeks depending on species, variety, and ecotype.

Seeds must be fully imbibed and not allowed to dry out for the entire treatment period. Keep the volume of seed per stratification bag relatively small to insure good aeration throughout. Placing the bags on wire mesh racks insures air exchange under the bag, and some nurseries hang the stratification bags from hooks. It is also a good practice to move and massage the bags weekly to move seeds in the interior to the outside and insure that no anaerobic conditions exist.

Presowing treatments to overcome double dormancy
The seeds of some species are particularly difficult to propagate because the dormancy level varies in the embryo or is due to a combination of factors; this is known as double dormancy. Developing seed treatments for these species can be especially challenging. Nature has devised some intricate tricks to insure the survival of the species by making sure seeds will germinate over a period of time, and it remains to be seen whether these ecological adaptations can be culturally overcome.

Overcoming seed dormancy is one of those factors that make nursery work so interesting because it is more of an art than a science.

Warm, moist stratification has two cultural objectives: to soften a hard seed coat and to encourage the growth of an underdeveloped embryo. The seeds of some species germinate better when a warm period immediately precedes cold stratification. The requirements are basically the same as for cold stratification except that the temperatures are increased to 18 to 29°C. Seeds must be fully imbibed to benefit from warm stratification and must be packaged and handled to encourage good aeration. The warm, moist stratification treatment period usually takes between four and twelve weeks, although this varies considerably among species. Some nurseries imbibe the seeds and place them in plastic bags, just as they do for cold stratification, and then hang them inside a warm greenhouse or place them on a heated floor or bench in a vegetative propagation structure.

Seed cleansing

Seedborne pathogens can be a major cause of nursery disease. The most common symptom of seedborne disease is slow and variable germination and emergence, although many growers often attribute this to poor quality seed. Even if germination occurs normally, seedborne fungi can cause post-emergence damping-off, or cotyledon blight. Sometimes, the symptoms of a seedborne disease problem do not become evident until root rot shows up later in the season. To manage seed disease, it is important to carefully observe germination and monitor seed use efficiency. If a problem is suspected, conduct seed assays to identify the specific pathogens involved and then treat seedlots that are shown to be contaminated.

As already discussed, soaking seed in an aerated rinse is a standard procedure for preparing nondormant seeds for sowing or as a treatment prior to stratification. It has also been shown to be an effective way to control nursery diseases and reduce the use of toxic pesticides (Dumroese et al. 1990).

Chemical sterilants such as hydrogen peroxide have been used to sterilize seeds, although the concentration and timing are critical. Chlorine bleach (sodium hypochlorite) also has been used on seeds as a surface sterilant. Another sterilization technique is the hot water soak. Place seeds in nylon mesh bags, immerse in a tank of warm water (49 to 57°C) for fifteen to thirty minutes, and then cool in running tap water (Handreck and Black 1994). Aerated steam is an even safer way to apply heat because it does not leach the seeds. Spread the seeds on a mesh screen in an insulated chamber connected to an aerated steamer so that the steam is dry when it reaches the chamber. Temperatures reach the same range as the hot water method and the treatment lasts thirty minutes, but a period as short as ten to fifteen minutes also may be effective. At the end of the treatment, lower temperatures rapidly by evaporative cooling and allow the seeds to dry (Hartman et al. 1990). Although the best temperature/time treatment depends on the pathogen to be controlled and would have to be determined for different species, hot water or steam is a safe, simple, and ecologically friendly way to cleanse seeds.

Historically, fungicides were routinely applied to seeds to control diseases, especially damping-off. But the number of registered pesticides decreases each year.

Seed sowing
The process of sowing varies with the type and quality of the seed. It begins with the calculation of the sowing rate based on the results of germination tests tempered by past experience. The decision will depend on seed availability and cost, germination test results, type of container, labor costs, and available growing space. One seed can be sown per container for high quality seedlots, but two to six (or more) seeds per container are used for seedlots with low germination percentages. Very small seeds are difficult to handle, not only because of their size but because static electricity makes them clump together. They can be sown with a wet toothpick or metal pin. Three different sowing techniques have been used in forest and conservation nurseries:

Direct seeding. This is the placement of seeds directly into the growth container when they are ready to germinate and grow. (Seeds with a dormancy requirement such as cold stratification must be treated prior to the planned sowing date.) The success of direct seeding depends on the accuracy of the seed test information, the sowing technique, and conditions in the propagation environment. Laboratory germination tests are conducted under ideal environmental conditions which are significantly different from those in an operational nursery. So nursery managers and others attempting to propagate native plants must use their experience in adjusting for this discrepancy.

Planting germinants from stratification. This technique involves sowing pregerminated seed into the growth containers and is particularly suited for large seeds. It is most commonly used for seeds requiring cold, moist stratification but can also be used for seeds requiring warm-moist presowing treatment. It can even be used for seeds of *Acer spp.* or *Juniperus spp.* that require both. Germinant sowing is particularly useful for seedlots of variable quality or for those for which no germination test data are available. Germinant sowing insures that a live seed is placed in every container and the resultant seedlings are larger because they can begin to grow immediately.

The process is relatively simple. Soak seeds (fresh or from storage) in an aerated water tank for twenty-four to forty-eight hours to achieve full imbibition and cleanse their seed coats. Then place them either in cold or warm, moist stratification until they begin to germinate. Growers place larger seeds in plastic bags containing a moist medium such as peat moss or wet burlap to maintain high humidity, whereas smaller seeds are stratified naked in a tray where the germinating seeds are easier to locate. A variety of common trays have been used, including styrofoam meat trays or cake pans with clear plastic lids. Place the bags or trays under normal refrigeration at 1 to 2°C for cold stratification, or leave under a moist cover in a greenhouse for warm stratification.

Check the seeds every few days or weeks and sow as soon as the radicle begins to emerge. Larger seeds can be planted by hand but smaller ones require tweezers. Some growers like to prune the radicle of dominant tap-rooted species, such as *Quercus spp.*, prior to planting to insure a more fibrous root system. The placement of the seeds is very important: they must be placed on their sides or with the radicle extending downward. Poorly placed seeds will develop a crook in their stems which can become brittle and break when they become larger. The major disadvantage of the germinant technique is that sowing will be extended over several weeks or even months, depending of the degree of seed dormancy, and the resultant crop development will be uneven, with seedlings over a range of sizes and ages.

Transplanting emergents from seed trays. Because many seeds of native species are too small or fragile to be direct seeded or planted as germinants, the seed is sown into shallow trays to germinate and then the emerging seedlings ("emergents") are transplanted into the growth container. This technique is particularly popular for native species because of their complex dormancy requirements and small or irregularly shaped seeds (Finnerty and Hutton 1993).

Seeds requiring scarification must be treated before sowing; sow those that require cold, moist stratification in trays in the fall and then place them in a refrigerator or even outside in a sheltered location to undergo natural stratification. In the latter case, the seed trays must be irrigated periodically to prevent desiccation and protected against rodent predation. When the trays are brought into the greenhouse in the spring, the seeds germinate almost immediately. Some species may germinate better at lower temperatures. For example, the seeds of *Gaultheria spp.* are sown in seed flats in cold frames where the temperature is around 13°C (Date 1994).

To propagate small, sensitive seeds, fill shallow trays with about 5 cm of standard peat moss-vermiculite growing medium and tamp it until it is firm, but not compacted. Scatter larger seeds over the surface by hand; a salt shaker is effective for very small seed. Cover the seeds with a light application of a fine-textured mulch such as sand-blasting grit, then irrigate, place into a greenhouse, and mist lightly. When they reach the cotyledon stage and begin to grow their first set of primary leaves, they are ready for transplanting to the growth containers. The best age for transplanting varies by species. Species with large cotyledons, such as *Rhus spp.*, should be transplanted at the two-leaf stage, whereas those with smaller cotyledons should not be transplanted until the six- or eight-leaf stage (Emery 1988).

The traditional transplanting technique consists of working the emergents loose from the seed tray, making a dibble hole in the growing media in the growth container, placing the plant in the hole, and firming the soil or growing medium around the stem. Growers have developed an innovative tool for this task. It is a small steel rod with a sharpened, flattened V-notch in the tip. The top of the emergent is held by one hand and the bottom of the root is hooked with the notched tip of the transplanting tool. The root is pushed down into the soil or growing medium until the seedling is at the proper depth. Then, while still stabilizing the seedling, the hooked bottom of the root is cut off and the tool removed. This simple technique leaves the emergent transplanted without the possibility of a "J-root" or other deformation, so that roots can develop normally. Transplanting emergents requires some degree of skill but can be easily mastered with some simple training. The procedure is labor intensive, compared to direct seeding or sowing germinants, but an experienced worker can transplant up to two thousand emergents in an eight-hour day (Landis and Simonich 1984).

Seed coverings

The most complete and rapid germination usually occurs when seeds remain uncovered and are watered by a misting system that keeps the medium surface damp. When seeds are watered less frequently, a covering is helpful in promoting germination. Seed coverings also help hold the seed in contact with the growing media, and reduce the development of cryptograms including moss, algae, and liverworts. Many materials have been used to cover seeds, including many standard components of growing media such as shredded peat moss, vermiculite, and perlite. Organic materials work reasonably well although they do encourage the growth of cryptogams in the moist environment necessary to germinate seeds (Tinus and McDonald 1979). Therefore, inorganic materials such as coarse sand or granite grit have become more popular in recent years. Light-colored materials should always be used because the intense sunlight in a greenhouse can quickly cause temperatures to reach damaging levels when dark mulches are used. Because the germinating seed is very susceptible to fungal or bacterial attack, any material in direct contact with it obviously should be sterile.

Growers should try to maintain a slightly acid pH (5.0 to 6.0) around the germinating seed to discourage damping-off fungi. Some seed coverings are made of crushed seashells, which are 90% calcium carbonate, and some coarse sands can also be calcareous. Such materials can cause the pH around the germinating seed to be too high; always test any potential new material. Peat moss has the ideal pH and the large fibers can be milled to the proper size by forcing them through a 6.4 mm (0.25 in.) mesh screen (Emery 1988).

Any seed covering must be of uniform depth or seed germination and emergence will vary widely, but the ideal depth depends on the size of the seed and, to some extent, on the type of irrigation system. Most seeds are phototropic and will germinate more quickly when they are not covered too deeply. A good general rule is to cover seeds to a depth approximately twice their smallest diameter. Leave very small seeds uncovered and mist frequently or cover with a fine-textured material such as milled peat moss (Emery 1988).

Vegetative Propagation

Vegetative propagation is defined as the production of new plants that contain the exact genetic characteristics of the parent plant. Because only one parent is required, vegetative propagation is also known as asexual propagation. By avoiding the genetic recombination inherent in sexual reproduction and seed development, nursery managers can produce multiple "copies" of an individual parent plant or group of plants with similar genetic composition. Vegetative propagation techniques vary considerably in effort and cost.

Vegetative propagation has many advantages: achieving rapid multiplication of selected plant material in a short time, maintaining a high degree of plant quality and uniformity, avoiding complex seed dormancy problems, and overcoming a shortage of available seed. Vegetatively propagated plants can flower earlier and more frequently than those grown from seed. On the other hand, vegetative propagation costs more than growing from seed, is labor intensive, involves loss of genetic diversity, and has the potential for plagiotropism and reduced plant vigor.

In many cases, however, vegetative propagation is the only option. Many native plant species do not produce seed every year, and crops of species growing in the same ecotype do not necessarily produce crops in the same year. Even if they do, the crops can be widely scattered and may not occur in the desired seed zone, so it may be impossible to obtain enough seed for an outplanting project. Seed quality also can vary considerably from year to year. In other situations, pests have so damaged the flowers or seeds that propagation by seed is not an option. Plants that have unique genetic properties, such as resistance to an insect or disease, can be propagated vegetatively so that the desired genes are not lost in sexual recombination. Vegetative propagation is also used to maintain or even increase populations of rare or endangered species.

Once the decision has been made to propagate a plant vegetatively, the best method to use depends on several factors including characteristics of the plant, the type of propagation environment, and the skill of the propagator. For a more detailed discussion on basic vegetative propagation concepts and procedures, the reader is referred to horticultural texts including Hartmann et al. (1990) and MacDonald (1986).

Stem Cuttings

Rooted stem cuttings are the most widely used vegetative propagation technique. Poplars and willows have traditionally been produced from rooted cuttings because their very small seeds are hard to handle, do not store well, and are covered with fine hairs which resist imbibition. Most species root well without rooting hormones although some recalcitrant species do require this assistance.

Type of cuttings. Stem cuttings are divided into categories depending on the type of plant and maturity of the tissue (Hartmann et al. 1990). *Hardwood cuttings* consist of mature tissue of the previous season's growth and are collected during the dormant period. They are easy to prepare, they store and ship well, and many do not require hormones or special propagation environments, although they can benefit from them. *Semihardwood cuttings* are collected from the semi-lignified tissue of actively growing plants. They are easy to prepare but should be used soon after collection and placed in a greenhouse with mist or a rooting chamber. They usually require treatment with hormones. *Softwood or herbaceous cuttings* are collected from the soft succulent new shoots of woody plants during the growing season. They must be used immediately and rooting hormones are required. Because the cuttings are very sensitive to desiccation, a mist bed or rooting chamber is necessary.

To produce cuttings at the nursery, it is possible, by repeatedly cutting a plant back to the base, to produce stump sprouts, or "hedges," which are then maintained in the nursery to provide a steady supply of cuttings. Hedges are more productive if the parent plant is cut all the way back to the root collar. Juvenile plant material typically produces roots naturally, but more mature cuttings can be stimulated to produce roots with hormone treatments.

Collecting Cuttings. An understanding of some basic physiological processes, along with some common sense, is essential to successful rooting of cuttings, starting with collection. Make field collections as early in the day as possible and on days with little or no wind in order to minimize water stress. Cloudy or foggy days are ideal. Move container stock plants to a shady area for several days and irrigate prior to collection if possible. Cuttings must use stored photosynthate to generate callus tissue and initiate new roots and so it makes sense to collect them during the dormant period when food reserves are the greatest.

Knowledge of the origin of cuttings is just as important as it is for seeds to insure that the nursery stock is well adapted to its outplanting environment. For restoration projects, cuttings should always be collected on or near the outplanting site if possible. If genetic diversity is a concern, make a wide selection of material. If there will be sufficient long-term demand for a particular ecotype, an easily accessible supply of cuttings can be obtained from outdoor stool beds, hedging orchards, or stock plants established in containers. All of these options reduce collection costs and maintain control over the source and quality of cuttings.

Rootability of stem cuttings varies greatly with the season of the year, and the best collection time depends on the type of cutting desired. Hardwood cuttings are collected during the winter dormant season whereas semihardwood cuttings are collected during the growing season when the shoots are not rapidly expanding. The best time to collect cuttings also varies by species and even genotype.

The physiological condition of the donor plant affects rooting success; cuttings should only be collected from healthy plants growing in an ideal environment for that plant, or from well-fertilized parent plants in the nursery. The size of the cuttings depends on the plant species, the type of material available, and the size of the container to be used. Stem cuttings have an inherent polarity and will always produce shoots at the distal end (nearest the bud) and roots at the proximal end (nearest the main stem or root system) (Hartmann et al. 1990). To distinguish between the top and bottom of hardwood cuttings, cut the bottoms at an angle which not only insures that the cuttings are planted right side up, but makes them easier to stick. Cuttings from lateral branches may exhibit plagiotropism, i.e., the horizontal growth habit maintained by some lateral cuttings after they are rooted. Plagiotropism is much more troublesome in some species than in others.

To prevent the spread of decay, pruning shears and other tools should be kept clean and disinfected regularly. Household bleach is the most common disinfectant because it is cheap and readily available. Although they are more expensive, benzalkonium chlorides and hydrogen peroxide are just as effective and have no potentially damaging breakdown products (McClelland and Smith 1994).

Hardwood cuttings of some deciduous species are relatively easy to root and many do not even require rooting hormones. Willow, poplar, and red-osier dogwood are collected as long whips which are then cut into the proper length for sticking. If they are collected by hand, the basal cut is typically made just below a node, where roots form more readily. The bundles of cuttings are then secured with a rubber band and stored under refrigeration at 0 to 4.5°C to keep them dormant until they are needed. Some nurseries soak their cuttings in a surface sterilant or fungicide to retard storage molds and decay of the cut surfaces. Placing the bottoms of the cuttings in moist sawdust or wood shavings while in the cooler keeps the bottoms warmer than the tops, which promotes callusing and speeds up root initiation. Warm-temperature callusing is achieved by treating the cuttings with rooting hormones and storing them under relatively warm, moist conditions (18 to 21°C) for three to five weeks. Some species that are difficult to root are even placed over bottom heat. Excessive temperatures or callusing time, however, can lead to lower survival after planting (Hartmann et al. 1990). In warmer climates, many species do not require any special processing and hardwood cuttings can be planted immediately.

Semihardwood or softwood cuttings must be handled differently because they have leaves and desiccate much more rapidly. They are collected from late summer to late fall and immediately placed in plastic bags in coolers to retard moisture loss and prevent overheating.

Rooting Hormones. Treating cuttings with rooting hormones greatly increases the speed and uniformity of root development on most cuttings, and many different products are commercially available. Rooting compounds may contain different hormones, at different concentrations, and other supplemental materials such as fungicides and vitamins. This can be confusing to the novice because there is no standard nomenclature for the contents or their concentration. It is important to read the label closely, test the effectiveness of different applications, and consult with other growers if possible.

Rooting hormones are applied to cuttings in three ways. Dipping in hormone powder is quick and easy for single cuttings. Powder formulations are safer to use because the hormone concentration is preset. However, it is difficult to control the application rate because the amount that adheres to the cutting depends on size, surface texture, and moisture content. Dipping in liquid for a few seconds is a simple and fast technique and provides uniform results when cuttings are bundled during treatment. However, liquid products must be diluted to the proper concentration, which leaves room for error. Liquid soaking for two to twenty-four hours is a slower method, but provides a more uniform application rate.

Culturing Cuttings. Growers have developed several cultural techniques to encourage cuttings to form roots while minimizing moisture loss. Wounding involves removing the outer bark or making incisions at the bottom of the cutting. This not only exposes the cambial tissue to the rooting hormone but encourages the formation of callus tissue, which is often the precursor of root initials. A moisture retardant like pruning sealant is sometimes applied to the exposed end of hardwood cuttings. Other cultural practices, such as removal of shoot tips or clipping or completely removing basal leaves, have been shown to improve rooting success with semihardwood or softwood cuttings.

Cuttings can be "stuck" or "set" into containers in two ways which differ not only in technique but with the type of cutting and the propagation facilities that will be required.

Direct sticking involves planting cuttings directly into growth containers and then moving them immediately to the propagation area where they will grow to desirable size. This technique is commonly used with hardwood cuttings of willows and cottonwoods, which root easily, or with semihardwood or softwood cuttings of species that require treatment with rooting hormones. Hardwood cuttings stick easily into the filled containers but less woody cuttings need a hole prepared in the container. Direct stick cuttings have simple cultural treatments, requiring only frequent irrigation or misting to minimize transpiration demand.

Prerooting is recommended for semihardwood and softwood cuttings that are more difficult to root, and treatment with rooting hormones and a special propagation environment are essential. Plant the treated cuttings into a shallow tray or small container until they begin to form roots which, depending on the species, can take from several weeks to many months. Determine root development by gently tugging the cutting—any resistance means that roots have formed and the cutting can be transplanted. Transplant the rooted cuttings into a container where they are cultured to desirable size. The newly formed roots are very tender and susceptible to breakage and so rooted cuttings must be carefully removed from the tray. Just as with other types of transplants, plant them carefully to avoid forming a "J-root." Irrigate transplanted cuttings frequently in the weeks following transplanting, and begin fertilization as soon as the plants have rooted.

The height of the trays and the characteristics of the media are critical to rooting success. Taller trays will provide better drainage than shallower ones. Rooting media should provide four functions: prevent disease, hold the cutting in place, supply water as fast as it is lost through transpiration, and permit easy penetration of air to the bottom of the cuttings. Cuttings are very susceptible to rot fungi and therefore the medium should be pasteurized. Fine sand or mixtures of perlite, vermiculite, peat, or coir have been used.

Growers should pay special attention to irrigation and temperature. Water loss is the single most important limiting factor, because cuttings do not have any roots to replenish moisture lost through transpiration. Some growers construct relatively simple rooting chambers or beds with bottom heat and humidity controls. Nurseries that produce a large number of cuttings use sophisticated propagation facilities that have completely controlled environments including automated fogging systems. Some nurseries have specially constructed chambers that generate a periodic mist or fog that keeps the relative humidity at almost 100%. Because so much water is required with mist nozzles or fog systems, the quality of irrigation water is critically important. Filter all irrigation sources to remove sediments that may plug nozzles; water high in dissolved salts will leave damaging and unsightly deposits on the cuttings. Both air and rooting medium temperatures are very important. Daytime air temperatures of 21 to 26°C are recommended for many plants. For some species, rooting media should also be kept at the same temperatures to promote rooting. Under-the-bench heat has proven to be effective in stimulating rooting. The effect of light on the rooting of cuttings is very complex as some species respond to different daylengths, intensity of light, and light quality. Most species do not generally require mineral nutrients until they are transplanted, although injecting a dilute nutrient solution into the misting system can replace the nutrients leached from the leaves.

Root Cuttings

Compared to stem cuttings, few species are propagated by root cuttings but this technique is possible with any plant that produces suckers in nature (Dirr and Heuser 1987). Collect sections from these roots, put them into trays filled with a moisture-holding growing medium, and place in a growth-promoting environment. After several weeks, shoots will begin to form on the root sections. Cut these shoots from the roots, treat with rooting hormones, and then stick into standard growth containers where they will root and form new plants. They are then cultured just like stem cuttings.

Layering

Layering involves forcing a section of stem or root into a favorable rooting environment so that it develops adventitious roots while still attached to the parent plant. The rooted section is then cut from the parent plant and transplanted into the growth container. Layering is a low-stress propagation method because the parent plant provides a steady supply of water and nutrients while roots are being formed. This is especially valuable when propagating threatened and endangered plants because there is little risk to the parent. Layering produces a relatively large plant in a short time and can be done without special propagation facilities. Because it is so labor intensive and the multiplication factor is low, layering is relatively expensive. Two types of layering have been used for forest and conservation plants.

Ground layering consists of bending a side shoot or branch over until it can be held in place and covered with growing medium or mulch. Rooting of the buried section is naturally stimulated by the interruption of the normal basipetal translocation of photosynthates so that they accumulate near the bend, and the exclusion of light. Cultural procedures that encourage rooting in stem cuttings, such as the use of hormones and wounding, also hasten the formation of roots.

In *air layering*, the bark is wounded or completely stripped from a section of stem or a lateral branch of the parent plant. The best success occurs on stems of the previous season's growth that still have mature leaves. Treat the wounded section with rooting hormone and immediately cover with damp peat moss or other moisture-holding material secured with a clear plastic wrap. Remove the protective wrapping periodically to check for root development and rewet the peat moss. The new plant is ready to be cut off the parent and transplanted in about two to three months but may require up to two seasons for species that are difficult to propagate. Air layering has been used to propagate many tropical and subtropical plants.

Division

Some plants naturally spread laterally by forming new shoots, rhizomes (modified stems), or bulbs and so they can be easily propagated by separation (removing naturally detachable structures) or division (cutting the plant into sections). Division is useful for propagating wetland species used for restoration or constructed wetland projects. The initial collections are made from native stands but then donor plants can be started in large containers or trays in the nursery. When they have grown large enough, divide the plants into sections and transplant to new containers. With some species, even the smallest section will root but the size of the divisions is critical with some species. Division can be repeated several times during the growing season and the new transplants are then given normal culture until they are ready for harvest (Beagle and Justin 1993). During harvesting, new plantlets can be collected and used to start the next crop. Like other vegetative propagation techniques, separation and division are relatively labor intensive but provide a quick and sure way to produce wetland species.

Grafting
Grafting is the art of propagating plants that are difficult to raise by other vegetative techniques. There are several types of grafts but all involve physically binding two plant parts together so that they will bond and grow into a single individual. The scion, the upper half of the graft which will develop into the new shoot system, is bonded to the rootstock, which is an established plant of the same or a closely-related species. Grafting is a very precise technique requiring considerable training and experience, and the reader is referred to McDonald (1986) and Hartmann et al. (1990) for more specific information.

Micropropagation
The newest and most rapidly developing vegetative propagation technique is micropropagation. This involves a series of sterile laboratory techniques in which small sections of plant tissue are chemically stimulated to form multiple shoots and are then rooted. The resultant "explants" are transplanted to growth containers and raised under normal culture. Although there are several different types of micropropagation, they all start by excising a small piece of plant tissue, cleansing it of microorganisms, and putting it in an artificial medium in a test tube or small laboratory vessel. By manipulating the laboratory environment and supplying specific hormones and vitamins, the grower causes these explants to multiply and develop into numerous miniature plantlets. Obviously, this process is quite complicated and the reader is referred to Hartmann et al. (1990) and Dirr and Heuser (1987) for a complete and thorough discussion.

Salvage

Some native plant growers obtain various species by salvaging them from a site designated for construction or some other use that will destroy the existing plant life. After securing permission from the landowner, carefully dig up the plants and transport to the nursery or outplanting site. It is crucial to keep the roots moist during this process. Cool, cloudy days with very little wind are ideal for this procedure. If possible, it is best to keep the original soil around the root system intact in order to minmize water loss and damage to the root system. Plants should be planted as soon as possible after salvage from a site.

References and Suggested Reading

This chapter was adapted from *Seedling Propagation, Volume 6, The Container Tree Nursery Manual.* (Landis, T.D., Tinus, R.W., McDonald, S.E., and Barnett, J.P. Agric. Handbk. 674., U.S.D.A., Forest Service., manuscript in progress) with generous permission from Tom Landis (Western Nursery Specialist, USDA Forest Service, Cooperative Forestry, P.O. Box 3623, Portland, OR 97208).

Forbs

Achillea millefolium
Yarrow

Description

Yarrow is a small, aromatic herbaceous perennial which grows 2-10 dm high. The erect stems are simple and sparse to dense villous. Leaves are alternate and lanceolate shaped with lanceolate-subulate segments. The blade is 3-15 cm long. The stems are topped with a cluster of small flowers which blossom throughout the summer. The flower head looks like a single blossom but is composed of ray and disk florets that constitute a complete flower. The ray florets are 2-4 mm long, three-toothed, with a pistillate, white, pink, or magenta. The disk florets are white to cream colored and consist of five lobes. Yarrow plants will form dense mats which can protect the soil from water and wind erosion. It is browsed by wildlife and livestock (Warwick and Black 1982, Strickler 1993).

Habitat and Geographic Range

Yarrow is generally found at elevations of 1200-3350 m but can also be found as low as sea level. Yarrow grows in meadows and pastures, along stream edges, in sand dunes, and along the edge of woodlots. It tends to grow on poor soils and is very drought tolerant (Warwick and Black 1982). Yarrow is distributed throughout the United States except Arizona, New Mexico, and parts of southern Texas (Hickerson 1986).

Propagation

Seed: Flowering starts in April at lower elevations continuing through September at higher elevations. Fruiting occurs during August and September. Collect seed heads and allow them to air dry indoors for at least a week before cleaning. Crush the heads with a rolling pin to remove the nutlets, then sieve the nutlets and chaff to separate the seed. Cold store dry seeds in a sealed container. Sow indoors in late winter and transfer to larger containers in six weeks, or sow outdoors in late April (Phillips 1985).

References

Bourdôt, G.W. 1984. Regeneration of yarrow (*Achillea millefolium* L.) rhizome fragments of different length from various depths in the soil. *Weed Research* 24:421-29.

Hickerson, J. 1986. *Achillea millefolium. In*: Fischer, William C. (comp.) The Fire Effects Information System [Monograph Online]. Missoula, MT: USDA Forest Service, Intermountain Fire Sciences Laboratory. http://www.fs.fed.us/database/feis/plants/Forb/ACHMIL. Accessed March 26, 1997.

Vegetative: Yarrow regenerates naturally via rhizomes. Plant rhizome fragments, 4 cm long (1-6 nodes), 5 cm deep; they will produce shoots in two to three weeks (Bourdôt 1984). Yarrow can also be propagated via division. Mature yarrow plants consist of a cluster of basal rosettes. Lift in early spring, then divide into individual or small groups of rosettes and replant. Cut back the leaves on each division to reduce moisture loss (Phillips 1985).

Phillips, H.R. 1985. *Growing and Propagating Wild Flowers.* Chapel Hill: University of North Carolina Press. 331p.

Strickler, D. 1993. *Wayside Wildflowers of the Pacific Northwest.* Columbia Falls, MT: Flower Press. 272p.

Warwick, S.I., and L. Black. 1982. The biology of Canadian weeds. 52. *Achillea millefolium* L. S.L. *Canadian Journal of Plant Science* 62: 163-82.

Anaphalis margaritacea
Pearly everlasting

References

Kruckeberg, A.R. 1982. *Gardening with Native Plants of the Pacific Northwest.* Seattle, WA: University of Washington Press. 252p.

Link, E. (ed.) 1993. Native Plant Propagation Techniques for National Parks: Interim Guide. East Lansing, MI: Rose Lake Plant Materials Center. 240p.

Peck, M.E. 1961. *A Manual of the Higher Plants of Oregon.* Portland, OR: Binfords & Mort Publishers. 936p.

Time Life, Inc. Electronic Encyclopedia. Virtual Garden. [Monograph Online]. http://pathfinder.com/@@yJuDF4F68wAAQAJ6/cgi-bin/VG/vg. Accessed June 26, 1996.

USDA Forest Service. 1988. *Range Plant Handbook.* New York: Dover Publications, Inc. 816p.

Wright, R.C.M. 1975. *The Complete Handbook of Plant Propagation.* New York: Macmillan Publishing Co., Inc. 191p.

Description
Pearly everlasting is an herbaceous perennial with many erect leafy stems up to 60 cm tall. The leaves are broadly to narrowly lance shaped, 5-15 cm long, alternate and stalkless. The leaves are white and wooly when young; the upper surface greens with age. The inflorescence is a terminal corymb composed of numerous small white flower heads. The bracts surrounding the flower head are involucre and papery, and persist indefinitely. The center of each head is yellow and composed of disk flowers with the ray flowers absent. Male and female flowers grow on different flower heads. The seed is a minute oblong achene with a tuft of fine bristles found only on the female plant (Peck 1961, Kruckeberg 1982, USDA 1988).

Habitat and Geographic Range
Pearly everlasting ranges from Newfoundland south to North Carolina, Kansas, west to California, and up into Alaska. It grows on open sunny sites, shaded hillsides, semi-dry slopes, banks of streams, mountain meadows, and basins from lowlands up into subalpine regions. It can also be found in burned-over and cut-over areas growing in dense stands (Kruckeberg 1982, USDA 1988).

Propagation
Seed: Seed matures from July to September. Collect by hand by clipping the heads, which can be dried by laying them out on tarps or in bags. Clean with a hammermill. Sow directly or in containers in the spring; planted very shallow (Wright 1975, Link 1993).
Seeds per kilogram: ~11,022,925-13,227,515 (Link 1993).
Vegetative: Dig up plants, divide, and replant when the clumps have become too dense, usually after three or four years of flowering. The best time of year for division is in the spring or fall (Wright 1975, Time Life, Inc. 1996).

Aquilegia formosa
Red columbine

Description
Red columbine is a perennial herb, 1.5-8 dm tall, arising from a simple to branched caudex. The leaves are mostly basal with a long petiole, the blades are triternate with each segment having two or three lobes. The leaves wither soon after the plant blooms. The flowers are bright red and yellow and are borne on stems sticking well above the foliage. The back of each petal is elongated into a long hollow spur containing nectar that attracts hummingbirds (Stead and Post 1989). The fruit is a follicle in groups of five (Helliwell 1987).

Habitat and Geographic Range
Red columbine is widely distributed in North America. It prefers moist woods and streamsides at elevations between 900 and 3100 m (Helliwell 1987, Link 1993).

Propagation
Seed: Red columbine blooms from May to August, depending on elevation and latitude. Most follicles dry and open at maturity. The tiny black seeds mature from June to August and can be harvested as soon as the seed heads are dry and come off easily by hand. Gently crush the dried heads to release the remaining seed, and then scalp with an air screen. Seeds can also be collected by cutting the fruiting stalk and placing in a bag before the follicles open. Dry the follicles in the bag for a few days and separate the seeds by shaking the bag. Seeds can be stored for up to two years at a low temperature and humidity or longer if stored in sealed containers in low moisture. Prechilling for three days is required for germination. Direct seed in spring or fall (Link 1993). Plant in containers or scatter evenly over a seedbed (this is made easier by first mixing the seeds with fine sand). Cover with a very thin layer of soil or weed-free compost and keep moist. Seeds should germinate in two to four weeks (Stead and Post 1989).
Seeds per kilogram: ~881,835 (Link 1993).

References
Helliwell, R. 1987. Forest Plants of the Warm Springs Indian Reservation. Warm Springs, OR: Confederated Tribes of the Warm Springs. 177p.

Link, E. (ed.) 1993. Native Plant Propagation Techniques for National Parks: Interim Guide. East Lansing, MI: Rose Lake Plant Materials Center. 240p.

Stead, S., and R.L. Post. 1989. Crimson columbine (*Aquilegia formosa*). Plants for the Lake Tahoe Basin. Soil Conservation Service, Nevada Cooperative Extension. Fact Sheet 89-52.

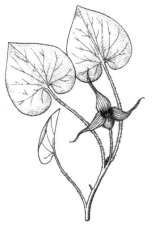

Asarum caudatum
Wild ginger

References

Pojar, J., and A. MacKinnon. 1994. *Plants of the Pacific Northwest Coast: Washington, Oregon, British Columbia, and Alaska*. Vancouver, BC, Canada: British Columbia Ministry of Forests and Lone Pine Publishing. 527p.

Strickler, D. 1993. *Wayside Wildflowers of the Pacific Northwest*. Columbia Falls, MT: Flower Press. 272p.

Time Life, Inc. Electronic Encyclopedia. Virtual Garden. [Monograph Online]. http://pathfinder.com/@@yJuDF4F68wAAQAJ6/cgi-bin/VG/vg. Accessed June 26, 1996.

Zimmerman, M.L,. and R.S. Griffith. 1991. *Asarum caudatum. In*: Fischer, William C. (comp.) The Fire Effects Information System [Monograph Online]. Missoula, MT: USDA, Forest Service, Intermountain Fire Sciences Laboratory. http://www.fs.fed.us/database/feis/plants/Forb/ASACAU. Accessed June 26, 1996.

Description

Wild ginger is a low-growing, trailing, perennial, evergreen herb that reaches 15-25 cm in height. The alternate, simple leaves are heart shaped, 5-15 cm across, prominently veined, aromatic, borne two per node, with long hairy petioles and entire margins. The flowers usually bloom near the ground and are often hidden by the leaves. They have three brownish-maroon, broad sepals that form a cup around the ovary and taper to long pointed tails. The fruit is a large capsule containing small angled or compressed seed (Zimmerman and Griffith 1991, Strickler 1993, Time Life, Inc. 1996). The roots can be ground and used as a ginger substitute (Pojar and MacKinnon 1994).

Habitat and Geographic Range

Wild ginger grows in moist, shaded woods below 1500 m and grows best in highly organic, slightly acidic soil. It ranges from British Columbia south to central California and is most commonly found from the Pacific Coast east to the Cascades. It can be found east of the Cascades, but less frequently, and is also found in areas of northern Idaho and western Montana (Zimmerman and Griffith 1991, Strickler 1993).

Propagation

Seed: Wild ginger flowers from April through July, depending on latitude and elevation (Zimmerman and Griffith 1991). The mature seed are difficult to collect but wild ginger seeds itself readily (Time Life, Inc. 1996).
Vegetative: Wild ginger has an extensive rhizome system. New plants can be made from divisions taken in early spring or fall when the parent plant is dormant. Plant the rhizomes 1 cm deep with the tip of the rhizome reaching the soil level. Space them 30 cm apart and keep the soil moist with a mulch of oak or beech leaves (Zimmerman and Griffith 1991, Time Life, Inc. 1996). Take root cuttings in the summer and start in sand for fall planting (Time Life, Inc. 1996).

Balsamorhiza sagittata
Arrowleaf balsamroot

Description
Arrowleaf balsamroot is a long-lived, cool-season, perennial forb. The basal leaves are large, often reaching 30 cm in length, arrow shaped, with entire margins and covered in dense, soft hairs that give them a silvery-gray appearance. The flowers are bright yellow, 7.5-13 cm across, and grow on leafless stems that reach 20-60 cm in height. The fruit is a glabrous achene. Arrowleaf balsamroot provides forage for many animals, and the seed is eaten by deer mice. It is a good species to use for revegetation of oil shale- or coal-mined lands and soil stabilization projects (Hermann 1966, Fischer and Holifield 1987, Tilford 1993).

Habitat and Geographic Range
Arrowleaf balsamroot ranges from British Columbia south to California and east to Saskatchewan, North Dakota, and Colorado. It ranges in elevation from 300 to 2900 m. It prefers well-drained, deep soils and open dry situations. It will tolerate semishade and can be found growing on open ridges, dry foothills, semiarid mountain rangelands, and southerly exposures (Hermann 1966, Fischer and Holifield 1987).

Propagation
Seed: Seasonal development varies due to geographical and elevational variation. Plants flower in May with seed ripening in mid-June and disseminating in late June through early August. Seed yield is generally abundant, but can be lost to late frosts, insects, and grazing animals. Viability of seed is often low due to insect damage. Harvest seed by hand or with a combine if terrain permits. Clean by drying, fanning, macerating, and fanning (Plummer et al. 1968). Seed can be stored at 20°C for up to five years. A cool, moist stratification for eight to twelve weeks at 0-4°C is required to break dormancy. Broadcast sow or drill in a firm seedbed, and cover following planting. Fall or winter sowing is recommended (Redente et al. 1982, Fischer and Holifield 1987).
Seeds per kilogram: ~121,790 (Plummer et al. 1968).

References
Fischer, W.C., and J.L. Holifield. 1987. *Balsamorhiza sagittata. In*: Fischer, William C. (comp.) The Fire Effects Information System [Monograph Online]. Missoula, MT: USDA Forest Service, Intermountain Fire Sciences Laboratory. http://www.fs.fed.us/database/feis/plants/Forb/ BALSAG. Accessed February 3, 1997.

Hermann, F.J. 1966. Notes on Western Range Forbs: Cruciferae through Compositae. USDA Forest Service Handbook No. 293. Washington D.C.: U.S. Government Printing Office. 365p.

Plummer, A.P., D.R. Christensen, and S.B. Monsen. 1968. Restoring big game range in Utah. Utah Div. Fish Game Publ. 68-3. 183p.

Redente, E.F., P.R. Ogle, and N.E. Hargis. 1982. Growing Colorado Plants from Seed: A State of the Art. Vol. III: Forbs. USDI Fish and Wildlife Service FWS/OBS-82/30. 141p.

Tilford, G.L. 1993. The *EcoHerbalist's Fieldbook: Wildcrafting in the Mountain West*. Conner, MT: Mountain Weed Publishing. 295p.

Vegetative: Arrowleaf balsamroot can regenerate vegetatively from the large, deep-seated, woody taproot topped by a many-headed caudex bearing rosettes of leaves from which new aerial stems arise. However, it is unknown whether new taproots are formed enabling propagation of individual plants (Fischer and Holifield 1987). Plants that are located in a construction area can be salvaged by transplanting the root crowns. The root can be dug up and the root crowns pulled laterally away from the root as carefully as possible. These should be transplanted to a similar environment (Tilford 1993).

Castilleja miniata
Common red paintbrush

Description

Common red paintbrush is an herbaceous perennial with few to several erect to ascending stems reaching 20-80 cm in height from a woody base. The leaves are alternate, narrow, linear to lance shaped with a sharp point, usually entire, but occasionally the upper leaves have three shallow lobes. The flowers are found in the axils of showy bracts. The four scarlet sepals are fused at the base. The petals are greenish, tubular, and fused into a narrow base. The fruit is a two-celled capsule (Pojar and MacKinnon 1994, Taylor and Douglas 1995).

Habitat and Geographic Range

Common red paintbrush is found throughout the Pacific Northwest growing from low to subalpine elevations. It can be found in open meadows and woods, dry prairies, stream banks, tidal marshes, thickets, grassy slopes, and shady forests (Pojar and MacKinnon 1994, Taylor and Douglas 1995).

Propagation

Seed: Seed kept in dry storage is viable for only a year and must be subjected to cool, moist conditions (1-5°C) for one to three months before it will germinate. Sown seeds will germinate at 21°C. Paintbrush is slow growing and susceptible to damping-off diseases. It should be fertilized with a constant supply of nitrogen. Seeds germinate best in the proximity of roots of other plants which they then parasitize (Borland 1994).
Vegetative: Paintbrush should behave like the figwort family and root from young shoots (Borland 1994).

References

Borland, J. 1994. Growing Indian Paintbrush. *American Nurseryman* 179(6):48-50, 52-53.

Pojar, J., and A. MacKinnon. 1994. *Plants of the Pacific Northwest Coast: Washington, Oregon, British Columbia, and Alaska.* Vancouver, BC, Canada: British Columbia Ministry of Forests and Lone Pine Publishing. 527p.

Taylor, R.J., and G.W. Douglas. 1995. *Mountain Plants of the Pacific Northwest: A Field Guide to Washington, Western British Columbia, and Southeastern Alaska.* Missoula, MT: Mountain Press Publishing Company. 437p.

Chimaphila umbellata
Prince's pine

References

Helliwell, R. 1987. Forest Plants of the Warm Springs Indian Reservation. Warm Springs, OR: Confederated Tribes of the Warm Springs. 177p.

Matthews, R.F. 1994. *Chimaphila umbellata. In*: Fischer, William C. (comp.) The Fire Effects Information System [Monograph Online]. Missoula, MT: USDA Forest Service, Intermountain Fire Sciences Laboratory. http://www.fs.fed.us/database/feis/plants/Shrub/CHIUMB. Accessed June 25, 1996.

Time Life, Inc. Electronic Encyclopedia. Virtual Garden. [Monograph Online]. http://pathfinder.com/@@yJuDF4F68wAAQAJ6/cgi-bin/VG/vg. Accessed June 25, 1996.

Description

Prince's pine is a small, erect perennial herb that grows 10-30 cm tall. The leaves are 3-7 cm long, whorled, oblanceolate and have sharply serrated margins. The pinkish flowers are borne three to fifteen in a terminal cluster. Each flower has five rounded petals. The fruit is a five-celled capsule, 7 mm across, containing numerous minute seeds and persisting through winter. In some parts of its range Prince's pine is browsed by deer and elk during the winter. The leaves are used in making root beer (Helliwell 1987, Matthews 1994, Time Life, Inc. 1996).

Habitat and Geographic Range

Prince's pine ranges from Newfoundland to Alaska, south to California and Mexico, and east to New Mexico, Colorado, and South Dakota. In the eastern United States it ranges from Maine south to Georgia and west to Minnesota. Its elevational range is 300-3500 m. It grows on a variety of soils but is most commonly found on dry, well-drained, rocky, sandy, acidic soils. Prince's pine is a moderately shade-tolerant species and is frequently found growing in old-growth and climax forests of the Pacific Northwest (Matthews 1994).

Propagation

Seed: Prince's pine flowers from June to August (Matthews 1994).

Vegetative: Take stem cuttings during the summer and root in a sand and peat medium. Outplant rooted cuttings in late spring with a 15-20 cm spacing (Time Life, Inc. 1996). Prince's pine produces long, fast-growing rhizomes. Those that are near the soil surface are able to produce new shoots (Matthews 1994). Prince's pine can also be propagated by division of underground stems (Time Life, Inc. 1996).

Cornus canadensis
Bunchberry

Description
Bunchberry is a low-growing, perennial herb reaching only 5-20 cm in height. The leaves are alternate but appear as if growing in two whorls. The leaves are up to 8.5 cm in length, 5 cm in width, lanceolate to ovate in shape, and have lateral veins that are nearly parallel with the margin. The small white flowers are borne in dense heads, with four toothed sepals, four reflexed petals, four stamens, and a somewhat flattened stigma making the entire inflorescence resemble a single flower. The fruit is a small red drupe with a small stone often containing a single seed. The seeds of bunchberry are much smaller than those of other dogwoods. The fruits are eaten by upland gamebirds and songbirds (Hall and Sibley 1976, Helliwell 1987).

Habitat and Geographic Range
Bunchberry grows in cold, boglike soils that are rich in sphagnum moss and slightly acidic. It prefers moist, shaded woods at moderate to high elevations and can be found ranging from Alaska to Greenland, and extending into the mountains of California in the west, and south to the mountains of West Virginia in the east (Helliwell 1987, Time Life, Inc 1996).

Propagation
Seed: The fruit ripens in August; collect by shaking or stripping from the branches. Clean and separate the seeds by macerating in water and floating off the pulp. Seed germinate best when sown right after cleaning in the fall. Plant 0.6-1.3 cm deep in a 3:1 sand:peat medium and cover with a layer of sawdust or pine needle mulch over winter. Seed can be stored in sealed containers at 3-5°C for two to four years. Seed sown in spring require stratification for 30 to 60 days at 25°C followed by 120-150 days at 0.5°C. Seeds may take up to three years to germinate. The establishment of new bunchberry plants from seed is low due to low fruit set, low germination and survival rates, and slow early growth (Brinkman 1974, Crane 1989, Time Life, Inc. 1996).
Seeds per kilogram: ~130,070-169,755 (Brinkman 1974).

References
Brinkman, K.A. 1974. *Cornus* L. Dogwood. pp. 336-42 *In*: Schopmeyer, C.S. (tech. coor.), *Seeds of Woody Plants in the United States.* Washington, D.C.: USDA Forest Service Agric. Handbook 450. 883p.

Crane, M.F. 1989. *Cornus canadensis. In*: Fischer, William C. (comp.) The Fire Effects Information System [Monograph Online]. Missoula, MT: USDA Forest Service, Intermountain Fire Sciences Laboratory. http://www.fs.fed.us/database/feis/plants/Shrub/CORCAN. Accessed June 27, 1996.

Hall, I.V., and J.D. Sibley. 1976. The Biology of Canadian Weeds. 20. *Cornus canadensis* L. *Canadian Journal of Plant Science* 56:885-92.

Helliwell, R. 1987. Forest Plants of the Warm Springs Indian Reservation. Warm Springs, OR: Confederated Tribes of the Warm Springs. 177p.

Kruckeberg, A.R. 1982. *Gardening with Native Plants of the Pacific Northwest*. Seattle, WA: University of Washington Press. 252p.

Time Life, Inc. Electronic Encyclopedia. Virtual Garden. [Monograph Online]. http://pathfinder.com/@@yJuDF4F68wAAQAJ6/cgi-bin/VG/vg. Accessed June 27, 1996.

Vegetative: Although some new seedlings are established by seed, bunchberry is a clonal perennial that relies heavily on vegetative regeneration to maintain itself and spread. It responds vigorously to disturbance (Crane 1989). Bunchberry grows laterally in organic matter by rhizomes with several dormant buds. Clumps can be transplanted but better establishment is achieved with well-grown seedlings set out in pots (Kruckeberg 1982).

Dicentra formosa
Pacific bleeding heart

Description

Pacific bleeding heart is an herbaceous perennial growing 20-45 cm tall. The leaves are basal with long petioles, biternately compound, fleshy textured, deeply cut, and smoothly glaucous. The pink to purple flowers are pendant, heart shaped, with deeply saccate or spurred petals, and are found coupled on stems rising above the foliage. The fruit is a two-valved capsule containing shiny black seeds (Schmidt 1980, Kruckeberg 1982).

Habitat and Geographic Range

Pacific bleeding heart ranges from southern British Columbia south to central California and is found mostly on the west side of the Cascade mountains. It grows at elevations up to 2100 m and prefers shady, moist, well-drained areas rich in organic matter (Schmidt 1980, Strickler 1993, Time Life, Inc. 1996).

Propagation

Seed: Seed matures during August and early September and is easily collected. Clean with a hammermill and low air flow. Sow seed fresh in late summer or fall. Link (1993) had 0% germination with seeds that were stored cool and dry, then cold stratified for 48 days. On the other hand, Buis (1996) kept the seed in a refrigerator over winter and obtained 50% germination with spring planting.

Vegetative: Pacific bleeding heart produces large rhizomes and is easily propagated by dividing the rhizomes after the plant has flowered (Time Life, Inc. 1996) or in early spring before flowering (Buis 1996). In addition, plants can easily be salvaged from sites designated for construction or other disturbance to plant life.

References

Buis, S. 1996. Owner, Sound Native Plants, Olympia, WA. Personal communication.

Kruckeberg, A.R. 1982. *Gardening with Native Plants of the Pacific Northwest.* Seattle, WA: University of Washington Press. 252p.

Link, E. (ed.) 1993. Native Plant Propagation Techniques for National Parks: Interim Guide. East Lansing, MI: Rose Lake Plant Materials Center. 240p.

Schmidt, M.G. 1980. *Growing California Native Plants.* Berkeley, CA: University of California Press. 366p.

Strickler, D. 1993. *Wayside Wildflowers of the Pacific Northwest.* Columbia Falls, MT: Flower Press. 272p.

Time Life, Inc. Electronic Encyclopedia. Virtual Garden. [Monograph Online]. http://pathfinder.com/@@yJuDF4F68wAAQAJ6/cgi-bin/VG/vg. Accessed June 25, 1996.

Epilobium angustifolium
Fireweed

References

Broderick, D.H. 1990. The biology of Canadian weeds. 93. *Epilobium angustifolium* L. (Onagraceae). *Canadian Journal of Plant Science* 70: 247-59.

Pavek, D.S. 1992. *Epilobium angustifolium*. *In*: Fischer, William C. (comp.) The Fire Effects Information System [Monograph Online]. Missoula, MT: USDA Forest Service, Intermountain Fire Sciences Laboratory. http://www.fs.fed.us/database/feis/plants/Forb/ EPIANG. Accessed February 3, 1997.

Time Life, Inc. Electronic Encyclopedia. Virtual Garden. [Monograph Online]. http://pathfinder.com/@@yJuDF4F68wAAQAJ6/cgi-bin/VG/vg. Accessed February 3, 1997.

Description

Fireweed is an herbaceous perennial with erect leafy stems that can reach over 2 m in height. It has a fine root system and rhizomes that can extend 45 cm into the soil although most grow only 0-15 cm deep. The leaves are alternate, entire, acuminate, and 3-20 cm long. The racemes are elongate, bracted, and bear fifteen or more pink flowers. The fruit is a capsule, 2.5-8 cm long, containing up to five hundred small, light brown seeds. The seeds have a tuft of long hairs at one end. The species is used by bees and is reported to be good sheep forage (Broderick 1990, Pavek 1992, Whitson et al. 1992).

Habitat and Geographic Range

Fireweed is circumboreal and is found in all of the Canadian provinces and throughout the United States except the southeastern states and Texas. It ranges from sea level to high alpine elevations, occurring most commonly on disturbed sites such as cut-over or burned timberland and swamps, avalanche areas, riverbars, along roadsides, in old fields, and on wasteland. It grows in a variety of soils from clays and clay-loams to sandy loams to unweathered parent material (Pavek 1992, Whitson et al. 1992).

Propagation

Seed: The seasonal development of fireweed varies greatly due to its wide geographic and ecological ranges. Flowers bloom from June through September with fruit maturing approximately one month later. Fireweed is a prolific seed producer. Seed is released starting in August and continuing after the shoots have died from frost injury. Seed can remain viable when stored for 18 to 24 months and can be sown in either spring or fall. Fall-sown seedlings will overwinter as a rosette. Germination is best if seeds are sown on the soil surface under warm, well-lighted, humid conditions. Germination is enhanced with added nutrients and reduced soil acidity (Broderick 1990, Pavek 1992, Time Life, Inc. 1997).

Vegetative: Rapidly growing shoots can readily sprout from fireweed rhizomes following disturbance and may bloom within one month (Pavek 1992). Fireweed can be propagated by root cuttings. Cuttings 32 cm long planted 5 cm deep had the highest success of shoot production, though shorter cuttings can be used (Broderick 1990).

Whitson, T.D. (ed.), L.C. Burrill, S.A. Dewey, D.W. Cudney, B.E. Nelson, R.D. Lee, and R. Parker. 1992. *Weeds of the West*. Newark, CA: Western Society of Weed Science. 630p.

Eriogonum nudum
Barestem buckwheat

Description

Barestem buckwheat is a perennial with stems that are smooth and slender and grow up to 90 cm tall. The basal leaves originate from a short woody crown. The leaves are slender stalked with dense, white woolly beneath turning nearly hairless on the upper surface. The flowers are usually white but occasionally tinted with rose or yellow, and are clustered on a repeatedly two- or three-forked inflorescence. The fruit is a hard dry achene that usually contains one seed. The young stems are eaten by livestock (Dayton 1960, Schmidt 1980).

Habitat and Geographic Range

Barestem buckwheat ranges from Washington south to California and Nevada. It prefers rocky, sandy, well-drained soils and usually grows in exposed, sunny, warm sites on dry hills, valley flats, and mountain slopes (Dayton 1960, USDA 1988).

Propagation

Seed: Barestem buckwheat flowers during July and August. Seed will germinate without pretreatment. Plant in the fall in a coarse soil medium and cover with sphagnum moss to help prevent damping-off. The following spring, transplant seedlings into 3-inch pots and allow to grow until they are large enough for outplanting (Schmidt 1980).

References

Dayton, W.A. 1960. Notes on Western Range Forbs: Equisetaceae through Fumariaceae. USDA Forest Service Handbook No. 161. Washington D.C.: U.S. Government Printing Office. 71p.

Schmidt, M.G. 1980. *Growing California Native Plants.* Berkeley, CA: University of California Press. 366p.

USDA Forest Service. 1988. *Range Plant Handbook.* New York: Dover Publications, Inc. 816p.

Eriogonum umbellatum
Sulphur buckwheat

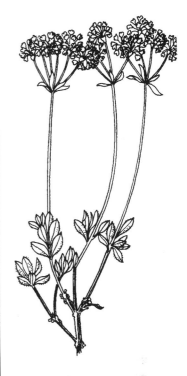

Description

Sulphur buckwheat is a low-growing, erect perennial with a tufted, woody crown and root. The leaves are basal or clustered at the ends of the branches and have dense white pubescence underneath. The flowers are sulphur yellow in color, hairless, with the calyx narrowed into a long, stalklike base. The involucres are long stalked, with eight bent-back lobes, in a simple umbel with a whorl of leaflike bracts at the base. The fruit is a hard, dry, usually one-seeded achene (Dayton 1960, Schmidt 1980, USDA 1988). The seeds are eaten by many species of birds and small mammals and the leaves by quail, deer, grouse, and mountain sheep (Stead and Post 1989).

Habitat and Geographic Range

Sulphur buckwheat grows on dry open sites in valleys and on mountain slopes from sea level up to subalpine elevations. It is found predominantly in dry areas and prefers rocky, sandy, and well-drained soils in places with moderate or low rainfall. It can be found from southern British Columbia south to California, and eastward to Colorado, Wyoming, and Montana (Dayton 1960, USDA 1988).

Propagation

Seed: Sulphur buckwheat flowers from June to August. Collect seed by manually rubbing the dried flower heads (Stead and Post 1989). It requires no pretreatment and can be planted in the fall in a coarse soil medium and covered with sphagnum moss to help prevent damping-off. Transplant seedlings into 3-inch pots the following spring and allow to grow until they are large enough to outplant (Schmidt 1980).

References

Dayton, W.A. 1960. Notes on Western Range Forbs: Equisetaceae through Fumariaceae. USDA Forest Service Handbook No. 161. Washington D.C.: U.S. Government Printing Office. 71p.

Schmidt, M.G. 1980. *Growing California Native Plants.* Berkeley, CA: University of California Press. 366p.

Stead, S., and R.L. Post. 1989. Sulfur flower, buckwheat (*Eriogonum umbellatum*). Plants for the Lake Tahoe Basin. Soil Conservation Service, Nevada Cooperative Extension. Fact Sheet 89-71.

USDA Forest Service. 1988. *Range Plant Handbook.* New York: Dover Publications, Inc. 816p.

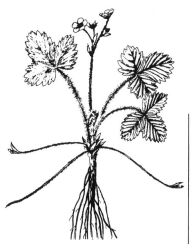

Fragaria vesca
Woodland strawberry

Description
Woodland strawberry is a perennial herb that grows up to 20 cm tall with long runners that creep along the ground. The yellowish-green oval leaves are palmately compound with three leaflets. The leaflets have coarsely serrated margins and long silky hairs on the surface, and are found at the end of a leafstalk. The flowers have five white petals, numerous yellow stamens, and grow on a separate shorter stalk than the leaves. The fruit is a red berry up to 2 cm long (Helliwell 1987, Snyder 1991) that is eaten by numerous mammals.

Habitat and Geographic Range
Woodland strawberry grows on cut-over lands, in roadside ditches, and in open woodland and rocky places. It prefers moderately drained sites and grows in full sun or light shade from low to subalpine elevations (Kruckeberg 1982, Govt. of Saskatchewan 1989, Snyder 1991, Pojar and Mackinnon 1994).

Propagation
Seed: The fruit of woodland strawberry ripens by late June. Clean the seed by macerating with water for twelve seconds and allowing it to sit until the seed falls and the pulp floats to the top. Seed can be stored and will remain viable for up to twenty years. A two- to three-month cold treatment and exposure to light will increase gemination. Plant in flats indoors six to eight weeks before the last frost or outdoors early in the spring; seed can take up to several months to germinate. Transplant six weeks after germination, allow to harden off, then set out in the field after six more weeks (Foster 1977, Time Life, Inc. 1996).

References
Buis, S. 1996. Owner, Sound Native Plants, Olympia, WA. Personal communication.

Foster, C.O. 1977. *Plants-a-Plenty: How to Multiply Outdoor and Indoor Plants Through Cuttings, Crown and Root Divisions, Grafting, Layering, and Seeds.* Emmaus, PA: Rodale Press, Inc. 328p.

Government of Saskatchewan. 1989. Guide to Forest Understory Vegetation in Saskatchewan. Canada Forestry. Technical Bulletin No. 9/1980 revised January 1989. 106p.

Helliwell, R. 1987. Forest Plants of the Warm Springs Indian Reservation. Warm Springs, OR: Confederated Tribes of the Warm Springs. 177p.

Vegetative: Cut newly rooted runners from the parent plant and transplant in early spring. Plant the crown even to the surface of the soil (Snyder 1991). Cut runners with at least two nodes from the plant during spring to midsummer. Treat the basal end with a mild-strength rooting hormone and stick into a growing medium of perlite:vermiculite (1:1). Place on a mist bench at 20°C. Once rooted, transplant into regular potting soil and place in heavy shade in a greenhouse (Buis 1996).

Kruckeberg, A.R. 1982. *Gardening with Native Plants of the Pacific Northwest*. Seattle, WA: University of Washington Press. 252p.

Pojar, J., and A. MacKinnon. 1994. *Plants of the Pacific Northwest Coast: Washington, Oregon, British Columbia, and Alaska*. Vancouver, BC, Canada: British Columbia Ministry of Forests and Lone Pine Publishing. 527p.

Snyder, L.C. 1991. *Native Plants for Northern Gardens*. Anderson Horticultural Library, University of Minnesota. 277p.

Time Life, Inc. Electronic Encyclopedia. Virtual Garden. [Monograph Online]. http://pathfinder.com/@@yJuDF4F68wAAQAJ6/cgi-bin/VG/vg. Accessed June 25, 1996.

Fragaria virginiana
Broadpetal strawberry

References

Buis, S. 1996. Owner, Sound Native Plants, Olympia, WA. Personal communication.

Foster, C.O. 1977. *Plants-a-Plenty: How to Multiply Outdoor and Indoor Plants Through Cuttings, Crown and Root Divisions, Grafting, Layering, and Seeds.* Emmaus, PA: Rodale Press, Inc. 328p.

Government of Saskatchewan. 1989. Guide to Forest Understory Vegetation in Saskatchewan. Canada Forestry. Technical Bulletin No. 9/1980 revised January, 1989. 106p.

Helliwell, R. 1987. Forest Plants of the Warm Springs Indian Reservation. Warm Springs, OR: Confederated Tribes of the Warm Springs. 177p.

Kruckeberg, A.R. 1982. *Gardening with Native Plants of the Pacific Northwest.* Seattle, WA: University of Washington Press. 252p.

Time Life, Inc. Electronic Encyclopedia. Virtual Garden. [Monograph Online]. http://pathfinder.com/@@yJuDF4F68wAAQAJ6/cgi-bin/VG/vg. Accessed June 25, 1996.

Description

Broadpetal strawberry is a low-growing perennial herb with long runners that creep along the ground. The bluish-green oval leaves are glaucous, glabrous, borne in threes with coarsely serrated margins and are found at the end of a leafstalk. The flowers are in groups of two to fifteen, with five white petals and numerous yellow stamens, and grow on a separate shorter stalk than the leaves. The fruit is a red berry that is smaller and more conical than cultivated strawberries (Helliwell 1987, Govt. of Saskatchewan 1989, Time Life, Inc. 1996). The berries are eaten by numerous mammals.

Habitat and Geographic Range

Broadpetal strawberry grows among grasses in open woods, fields, and meadows. It does best in well to imperfectly drained humus-filled soils and is usually found at low to mid-elevations, but occasionally grows at high elevations (Kruckeberg 1982, Helliwell 1987, Govt. of Saskatchewan 1989, Time Life, Inc. 1996).

Propagation

Seed: The fruit ripens by late June. Clean by macerating for twelve seconds and allowing the pulp to float off. Seed can be stored and will remain viable for up to twenty years. Germination is increased with a two- to three-month cold treatment and exposure to light. Planted in flats; seed may take several months to germinate. Transplant six weeks after germination, allow to harden off, then set out in the field after six more weeks (Foster 1977).

Vegetative: Cut newly rooted runners from the parent plant and transplant in late summer or early fall. Keep the crown above the surface of the soil (Foster 1977, Time Life, Inc. 1996). Cut runners with at least two nodes from the plant during spring to midsummer. Treat the basal end with a mild-strength rooting hormone and stick into a growing medium of perlite:vermiculite (1:1). Place on a mist bench at 20°C. Once rooted, plants can be transplanted into regular potting soil and placed in heavy shade in a greenhouse (Buis 1996).

Heracleum lanatum
Cow-parsnip

Description

Cow-parsnip is a perennial forming a low-growing rosette, with a large, fleshy taproot its first year. The stems grow 1-3 m and are jointed, pubescent, and hollow. Leaves are divided into three large, irregularly-toothed leaflets that are hairy and 10-25 cm wide. The cream-colored flowers are found in clusters with five petals in the umbels, the outer petals being enlarged. Seeds are ovate shaped, have ridges that alternate with four black lines and are flat on one side and round on the other. Cow-parsnip is a forage plant for deer, elk, moose, and bear. It is good for erosion control and both short- and long-term revegetation projects (Hermann 1966, Whitson et al. 1992, Tilford 1993, Esser 1995).

Habitat and Geographic Range

Cow-parsnip ranges from Alaska to Newfoundland south to California, Arizona, and Georgia. It ranges in elevation from sea level to 3750 m. It can be found in woodlands, forest openings, riparian areas, and grasslands. It grows best on moist loam and sandy loam soils with good drainage. Cow-parsnip is shade tolerant, but can also be found growing in some open habitats. It is considered a seral and climax community species and is commonly found as an understory species in quaking aspen and red alder community types (Hermann 1966, Esser 1995).

Propagation

Seed: Cow-parsnip flowers from May through August depending on its geographic and elevational range. Collect seed by hand-stripping into a container in late summer after the dark stripes are prominent. Clean by drying and fanning (Plummer et al. 1968). Seed can be stored in airtight containers at 2°C for up to three years. No stratification is required but seed should be leached in running tap water for four hours before planting. Fluctuating diurnal temperatures of 22°C (for 8 hours) and 17°C (for 16 hours) are recommended for germination (Redente et al. 1982). Seed can be broadcast or drill-seeded in the fall at 1-2 kilograms per hectare (Esser 1995).

Seeds per kilogram: ~167,550 (Redente et al. 1982).

References

Esser, L.L. 1995. *Heracleum lanatum. In*: Fischer, William C. (comp.) The Fire Effects Information System [Monograph Online]. Missoula, MT: USDA Forest Service, Intermountain Fire Sciences Laboratory. http://www.fs.fed.us/database/feis/plants/Forb/HERLAN. Accessed February 6, 1997.

Hermann, F.J. 1966. Notes on Western Range Forbs: Cruciferae through Compositae. USDA Forest Service Handbook No.293. Washington D.C.: U.S. Government Printing Office. 365p.

Plummer, A.P., D.R. Christensen, and S.B. Monsen. 1968. Restoring Big Game Range in Utah. Utah Div. Fish Game Publ. 68-3. 183p.

Redente, E.F., P.R. Ogle, and N.E. Hargis. 1982. Growing Colorado Plants from Seed: A State of the Art. Vol. III: Forbs. USDI Fish and Wildlife Service FWS/OBS-82/30. 141p.

Tilford, G.L. 1993. *The EcoHerbalist's Fieldbook: Wildcrafting in the Mountain West*. Conner, MT: Mountain Weed Publishing. 295p.

Whitson, T.D. (ed.), L.C. Burrill, S.A. Dewey, D.W. Cudney, B.E. Nelson, R.D. Lee, and R. Parker. 1992. *Weeds of the West*. Newark, CA: Western Society of Weed Science. 630p.

Vegetative: It is unclear whether cow-parsnip regenerates vegetatively. However, Esser (1995) noted that it had been included in a group of plants that regenerate rapidly from subsurface adventitious buds.

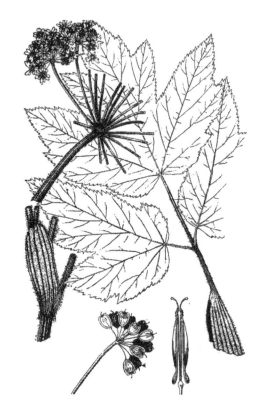

Hieracium albiflorum
White hawkweed

Description

White hawkweed is a perennial herb with solitary erect
stems growing 30-120 cm in height and containing a
milky juice. It has a fibrous root system growing from a
somewhat thick, short rootstock. The leaves are mostly
basal, with entire to wavy margins. Lower leaves are
stalked, oblong to reverse lance shaped and 5-15 cm
long. Stem leaves are alternate, linear to lance shaped,
stalkless, and smaller than the lower leaves. The flowers
are small, white, and borne fifteen to thirty in cymes.
The fruit is a reddish-brown, many-ribbed, 3-mm-long
achene with a dense tuft of pappus bristles. White
hawkweed is grazed by livestock, deer, and elk (USDA
1988, Pojar and MacKinnon 1994).

Habitat and Geographic Range

White hawkweed ranges from Alaska, east to
Saskatchewan, and south to Colorado and California. It
can be found from 100 to 2900 m. It grows in dry to
somewhat moist open forests, meadows, wooded slopes,
clearings, and roadsides. It is most commonly found
growing under ponderosa pine and lodgepole pine
(USDA 1988, Pojar and MacKinnon 1994).

Propagation

Seed: Collect seed in August, depending on geographic
location and elevation. Sow in the fall or moist stratify
for 90 days at 1°C and sow in the spring. Germination is
slow and may take three months to reach 50%
germinative capacity (Redente et al. 1982, Romme et al.
1995).

References

Pojar, J., and A.
MacKinnon. 1994. *Plants
of the Pacific Northwest
Coast: Washington,
Oregon, British Columbia,
and Alaska.* Vancouver,
BC, Canada: British
Columbia Ministry of
Forests and Lone Pine
Publishing. 527p.

Redente, E.F., P.R. Ogle,
and N.E. Hargis. 1982.
Growing Colorado Plants
from Seed: A State of the
Art. Vol. III: Forbs. USDI
Fish and Wildlife Service.
FWS/OBS-82/30. 141p.

Romme, W.H., L.
Bohland, C. Persichetty,
and T. Caruso. 1995.
Germination ecology of
some common forest
herbs in Yellowstone
National Park, Wyoming,
U.S.A. *Arctic and Alpine
Research* 27(4):407-12.

USDA Forest Service.
1988. *Range Plant
Handbook.* New York:
Dover Publications, Inc.
816p.

Ipomopsis aggregata
(Gilia aggregata)
Scarlet gilia

References

Art, H.W. 1990. *The Wildflower Gardener's Guide.* Pacific Northwest, Rocky Mountain, and Western Canada Edition. Pownal, VT: Storey Communications, Inc. 170p.

Hermann, F.J. 1966. Notes on Western Range Forbs: Cruciferae through Compositae. USDA Forest Service Handbook No.293. Washington, D.C.: U.S. Government Printing Office. 365p.

Parish, R., R. Coupe, and D. Lloyd. 1996. *Plants of Southern Interior British Columbia.* Vancouver, BC, Canada: British Columbia Ministry of Forests and Lone Pine Publishing. 463p.

Description

Scarlet gilia is a biennial or short-lived perennial growing 20-100 cm in height. It has one to several stems and is sticky and hairy on the upper areas. The pinnate leaves are highly dissected into narrow segments, dense near the base of the plant and sparse farther up the stem, 10 cm long, and emit a skunklike odor when crushed. The flowers are bright scarlet, 1.9-3.8 cm long, occur five to fifteen per stem, and are borne in a loose inflorescence. The corolla is a long tube with spreading lobes and the calyx is of short green ribs and transparent segments. The fruit is a capsule containing a dozen small seeds. Native Americans crushed the roots to create a blue dye and boiled the whole plant down to make a glue. Several compounds that appear to be effective in fighting cancer have been found in the leaves. Plants are eaten by mule deer, elk, and pocket gophers (Art 1990, Parish et al. 1996).

Habitat and Geographic Range

Scarlet gilia ranges from Montana to British Columbia and south to California, New Mexico and Texas. It grows at low to moderately high elevations. It is found on open, rocky slopes, grasslands, open forests, plains, hills, dry meadows, and mountain slopes (Hermann 1966, Parish et al. 1996).

Propagation

Seed: Scarlet gilia reproduces solely by seed. Plants start producing flowers from one to eight years of age. Plants flower from late spring to summer and fruit ripens in the summer. No pretreatment is required. Plant into flats and keep moist until the seedlings are established. The plant forms a basal rosette in its first year of growth (Art 1990).

Linnaea borealis
Twinflower

Description

Twinflower is a low, trailing, evergreen forb. Aerial stems arise from the stolon, which may become shallowly buried. The aerial stems become woody with age. Twinflower has a fibrous root system growing just beneath the soil surface. The leaves are oval-shaped, opposite, 12 mm long, 8-15 mm wide, with a slightly wavy margin. The flowers are pink, borne in pairs on an upright stalk which is 2-10 cm tall, and hang downward from the top of the stalk. The fruit is an oval capsule, 2-5 mm in diameter, and covered with short sticky hairs (Buis 1989, Government of Saskatchewan 1989, Howard 1993).

Habitat and Geographic Range

Twinflower ranges from polar regions in the northern hemisphere south to California, New Mexico and West Virginia. It grows under part shade with some organic matter in the soil, but can also be found in full sunlight. Twinflower grows in moist, dense forests to scattered wooded areas, shrub thickets, and rocky shorelines. It can be found from recently disturbed sites to climax plant communities. It grows at various elevations up to timberline (Buis 1989, Howard 1993, Strickler 1993, Pojar and MacKinnon 1994).

Propagation

Seed: Twinflower blooms from June to September with flowers lasting about seven days. The fruit matures approximately 36 days after flowering. Allow the seed to air dry, then plant in the fall. If planted in spring, stratify at 1°C for 60 days. One study showed little difference in germination capacity between refrigerated and unrefrigerated seed (Nichols 1934, McLean 1967, Redente et al. 1982, Howard 1993).

References

Buis, S. 1989. Propagation and use of twenty native Pacific Northwest plant species. Draft report for the National Park Service, unpublished. Olympia, WA.

Government of Saskatchewan. Forestry Canada. 1989. Guide to Forest Understory Vegetation in Saskatchewan. Technical Bulletin No. 9. 106p.

Howard, J.L. 1993. *Linnaea borealis. In*: Fischer, William C. (comp.) The Fire Effects Information System [Monograph Online]. Missoula, MT: USDA Forest Service, Intermountain Fire Sciences Laboratory. http://www.fs.fed.us/database/feis/plants/Shrub/LINBOR. Accessed February 3, 1997.

McLean, A. 1967. Germination of forest range species from southern British *Columbia Journal of Range Management* 20(5):321-22.

Nichols, G.E. 1934. The influence of exposure to winter temperatures upon seed germination in various native American plants. *Ecology* 15:364-73.

Pojar, J., and A. MacKinnon. 1994. *Plants of the Pacific Northwest Coast: Washington, Oregon, British Columbia, and Alaska*. Vancouver, BC, Canada: British Columbia Ministry of Forests and Lone Pine Publishing. 527p.

Redente, E.F., P.R. Ogle, and N.E. Hargis. 1982. Growing Colorado Plants from Seed: A State of the Art. Vol. III: Forbs. USDI Fish and Wildlife Service. FWS/OBS-82/30. 141p.

Strickler, Dr. D. 1993. *Wayside Wildflowers of the Pacific Northwest*. Columbia Falls, MT: The Flower Press. 272p.

Vegetative: Twinflower can be propagated from stem cuttings taken in spring or early summer. Stick cuttings 7-13 cm long in a peat:sand mixture (1:1) in a coldframe and keep moist. Rooting should occur in three to six weeks (Buis 1989). Stolons are first produced when plant is five to ten years of age (Howard 1993).

Lupinus latifolius
Broadleaf lupine

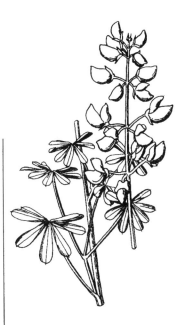

Description

Broadleaf lupine is a many-stemmed perennial herb that grows 0.5-1.5 m tall. The stems are erect with thinly appressed pubescence. The leaves are palmately compound with seven to nine oblanceolate-shaped leaflets, 4-8 cm long, that are dark green above and whitish underneath. The flowers are dark blue to purplish in color, pea-type with an unspurred calyx, and borne in dense, long pyramidal racemes. The fruit is a dark, flattened, sparsely hairy pod, 2-3.5 cm long, containing several seeds. The foliage may be somewhat palatable to livestock but the seeds are poisonous (Gilkey and Dennis 1980, Helliwell 1987).

Habitat and Geographic Range

Broadleaf lupine grows from lowland to subalpine elevations in the Cascade Mountains from British Columbia south to California, and west to the coast. It can be found in moist or open woods and along roadsides (Gilkey and Dennis 1980, Helliwell 1987, Link 1993).

Propagation

Seed: Lupine seed matures from June to September. Collect by hand while the pods are still closed and put into cloth bags for drying. Clean the seed with a hammermill and store in a cool dry environment for as long as several years. Pretreat the seed by soaking in hot water for one to 16 hours, then inoculate with *Rhizobium lupini* at the time of planting. Broadleaf lupine can be propagated by direct seeding or by growing plugs for transplant (Link 1993).
Seeds per kilogram: ~771,605 (Link 1993).
Vegetative: Take cuttings from side shoots of hardened stems in the spring (Foster 1977).

References

Foster, C.O. 1977. *Plants-a-Plenty: How to Multiply Outdoor and Indoor Plants Through Cuttings, Crown and Root Divisions, Grafting, Layering, and Seeds.* Emmaus, PA: Rodale Press, Inc. 328p.

Gilkey, H.M., and L.J. Dennis. 1980. *Handbook of Northwestern Plants.* Corvallis, OR: Oregon State University Bookstores, Inc. 507p.

Helliwell, R. 1987. Forest Plants of the Warm Springs Indian Reservation. Warm Springs, OR: Confederated Tribes of the Warm Springs. 177p.

Link, E. (ed.) 1993. Native Plant Propagation Techinques for National Parks: Interim Guide. East Lansing, MI: Rose Lake Plant Materials Center. 240p.

Lupinus lepidus
Prairie lupine

References

Foster, C.O. 1977. *Plants-a-Plenty: How to Multiply Outdoor and Indoor Plants Through Cuttings, Crown and Root Divisions, Grafting, Layering, and Seeds.* Emmaus, PA: Rodale Press, Inc. 328p.

Gilkey, H.M., and L.J. Dennis. 1980. *Handbook of Northwestern Plants.* Corvallis, OR: Oregon State University Bookstores, Inc. 507p.

Link, E. (ed.) 1993. Native Plant Propagation Techinques for National Parks: Interim Guide. East Lansing, MI: Rose Lake Plant Materials Center. 240p.

Schmidt, M.G. 1980. *Growing California Native Plants.* Berkeley, CA: University of California Press. 366p.

Strickler, D. 1993. *Wayside Wildflowers of the Pacific Northwest.* Columbia Falls, MT: Flower Press. 272p.

Description

Prairie lupine is a low, spreading, many-stemmed, decumbent perennial. There are brownish-yellow to silver appressed hairs over the whole plant. The floral stems reach 10-25 cm and grow among and above the leaves. The leaves are palmately compound with five to seven oblanceolate leaflets that are 1-4 cm long. Most of the leaves are basal but some originate on the stems. The dark blue to purple, sometimes white, flowers are borne in dense spikes. The flowers are 8-13 mm long with the banner petals bent sharply backward and usually a different color from the wings and keels. The fruit is a flattened pod containing two to twelve seeds (Gilkey and Dennis 1980, Strickler 1993).

Habitat and Geographic Range

Prairie lupine ranges from British Columbia south to California and east to Montana and Colorado. It grows in open prairies up into the mountains (Strickler 1993).

Propagation

Seed: Flowers bloom from June to August. Seed can be hand collected but collecting is slow due to the small size of the plant. Dry seed in the pods in paper bags, hand thresh, and screen to remove the pods and stem material. Store air-dried seed under cool, dry conditions. Seed requires scarification: shake in a jar half-filled with coarse sand, or scarify gently with sandpaper, soak in hot water until the water has cooled, then plant. It is best to sow seed directly since lupine do not do well when overhandled (Foster 1977, Schmidt 1980, Link 1993).
Vegetative: Take cuttings from side shoots of hardened stems in the spring (Foster 1977).

Lupinus sericeus
Silky lupine

Description

Silky lupine is a perennial, cool-season forb that grows 20-50 cm tall. Stems are simple or branched and arise from a woody caudex. The leaves are mostly basal with seven to nine leaflets. The leaflets are 3-7 cm long, narrow, acute, and densely subappressed and silky on both sides. The racemes are 12-15 cm long with blue to purple flowers on densely appressed-pubescent pedicels. The calyx is also densely silky. The fruit is a pod, 2.5-3 cm long, containing two to six seeds. Silky lupine is eaten by deer and upland game birds. It is also used as both food and cover by small birds and mammals. It is a good species for rehabilitation of disturbed sites due to its ability to fix nitrogen (Peck 1961, Link 1993, Matthews 1993).

Habitat and Geographic Range

Silky lupine ranges from British Columbia south through Washington and Oregon on the east side of the Cascade Mountains, to California and Arizona and east into New Mexico, Colorado, South Dakota, Montana, and Alberta. It grows up to 3000 m in elevation. Its habitat includes grasslands, sagebrush, mountain brush, and aspen and conifer forests. It can be found on dry, rocky sites with gentle to steep slopes but grows best in dry, sandy-loam and clayey-loam soils (Matthews 1993).

Propagation

Seed: Seed matures from early to late August. Collection can be time consuming due to the low number of seeds per pod. Clean seed by running dried pods through a hammermill and a fanning mill. Store in a cold, dry environment. An insecticide may be required to protect seed from the insect larvae found within it. Estimated storage time is five to ten years without much loss in viability. Link (1993) recommends inoculating the seed with a rhizobium before sowing. It is best to direct seed (in full sun to partial shade) since lupine does not tolerate much handling (Foster 1977, Link 1993, Matthews 1993).
Seeds per kilogram: ~43,430-92,595 (Link 1993).
Vegetative: Take cuttings from side shoots of hardened stems in the spring (Foster 1977).

References

Foster, C.O. 1977. *Plants-a-Plenty: How to Multiply Outdoor and Indoor Plants Through Cuttings, Crown and Root Divisions, Grafting, Layering, and Seeds*. Emmaus, PA: Rodale Press, Inc. 328p.

Link, E. (ed.) 1993. Native Plant Propagation Techinques for National Parks: Interim Guide. East Lansing, MI: Rose Lake Plant Materials Center. 240p.

Matthews, R.F. 1993. *Lupinus sericeus. In:* Fischer, William C. (comp.) The Fire Effects Information System [Monograph Online]. Missoula, MT: USDA Forest Service, Intermountain Fire Sciences Laboratory. http://www.fs.fed.us/database/feis/plants/Forb/LUPSER. Accessed September 11, 1996.

Peck, M.E. 1961. *A Manual of the Higher Plants of Oregon*. Portland, OR: Binford & Mort. 936p.

Osmorhiza occidentalis
Sweetanise

References

Peck, M.E. 1961. *A Manual of the Higher Plants of Oregon.* Portland, OR: Binfords & Mort Publishers. 936p.

Plummer, A.P., D.R. Christensen, and S.B. Monsen. 1968. Restoring Big Game Range in Utah. Utah Div. Fish Game Publ. 68-3. 183p.

Redente, E.F., P.R. Ogle, and N.E. Hargis. 1982. Growing Colorado Plants from Seed: A State of the Art. Vol. III: Forbs. USDI Fish and Wildlife Service. FWS/OBS-82/30. 141p.

Stevens, R., N. Shaw, and C.G. Howard. 1985. Important nonleguminous forbs for Intermountain ranges. pp. 210-20 *In*: Symposium on Range Plant Improvment in Western North America: Status and Future. Salt Lake City, UT; February 14, 1985.

USDA Forest Service. 1988. *Range Plant Handbook.* New York: Dover Publications, Inc. 816p.

Description

Sweetanise is a perennial herb with many stout, erect, hollow, glabrous stems that reach 6-12 dm in height. The roots are woody, and deep set. The leaves are bipinnate and slightly hairy. The leaflets are ovate to lanceolate, 4-8 cm long, strongly veined, with coarsely serrated margins. The small white to yellowish flowers are on short stalks that are borne on five to twelve erect longer stalks which form a compound umbel at the end of a stem. The fruit is a schizocarp that separates when ripe. The seed is linear, 12-17 mm long, five-angled, and noticeably ribbed. Sweetanise is grazed by livestock, deer, and elk (Peck 1961, USDA 1988).

Habitat and Geographic Range

Sweetanise ranges from Alberta and British Columbia south to California and Colorado. Its elevational range is 600-3050 m. It prefers well-drained soils and is most commonly found in cool, moist woods, forest openings, and valleys, and on brushy hillsides (Stevens et al. 1985, USDA 1988).

Propagation

Seed: Sweetanise flowers during May and June and seeds mature in August and September. Seed, collected by hand-pulling the heads, is then dried and cleaned with a fanning mill. It can be stored in airtight containers at 2°C for up to five years. Fluctuating diurnal temperatures of 22°C (for 8 hours) and 17°C (for 16 hours) are recommended for germination. Sweetanise can also be direct seeded in the fall at a depth of 6 mm (Plummer 1968, Redente et al. 1982, Stevens et al. 1985).
Seeds per kilogram: ~65,795 (Plummer 1968).

Oxalis oregana
Redwood sorrel

Description

Redwood sorrel, also called Oregon wood-sorrel, is a perennial with brownish, hairy, flowering stems growing 5-15 cm tall from a creeping rootstock. The cloverlike leaves are compound, with three to four palmately arranged leaflets. The leaflets are light sensitive and fold downward at night or in overcast weather. The pink or white flowers grow singly atop a long stalk, with five petals, five sepals, and ten stamens. The fruit is a football-shaped, five-chambered capsule, 7-9 mm long, that ruptures when ripe to distribute the almond-shaped seed (Pojar and MacKinnon 1994).

Habitat and Geographic Range

Redwood sorrel ranges from Vancouver Island, British Columbia, south along the coast to mid-California. It can be found in moist, forested areas at low to mid-elevations. It grows well in acid soils rich in organic matter (Art 1990, Pojar and MacKinnon 1994).

Propagation

Seed: Redwood sorrel flowers from late winter to late summer and fruits any time from spring to early fall. The seed requires no pretreatment and can be planted into flats and kept moist until seedlings are well established (Art 1990).

Vegetative: Redwood sorrel can be propagated by division. Make divisions from mature rhizomes in early spring and replant to a depth of 1.3 cm (Art 1990).

References

Art, H.W. 1990. *The Wildflower Gardener's Guide*. Pacific Northwest, Rocky Mountain, and Western Canada Edition. Pownal, VT: Storey Communications, Inc. 170p.

Pojar, J., and A. MacKinnon. 1994. *Plants of the Pacific Northwest Coast: Washington, Oregon, British Columbia, and Alaska*. Vancouver, BC, Canada: British Columbia Ministry of Forests and Lone Pine Publishing. 527p.

Penstemon procerus
Small-flowered penstemon

References

Pojar, J., and A. MacKinnon. 1994. *Plants of the Pacific Northwest Coast: Washington, Oregon, British Columbia, and Alaska*. Vancouver, BC, Canada: British Columbia Ministry of Forests and Lone Pine Publishing. 527p.

Prockter, N.J. 1976. *Simple Propagation: Propagating by Seed, Division, Layering, Cuttings, Budding and Grafting*. London: Faber and Faber. 246p.

Redente, E.F., P.R. Ogle, and N.E. Hargis. 1982. Growing Colorado Plants from Seed: A State of the Art. Vol. III: Forbs. USDI Fish and Wildlife Service. FWS/OBS-82/30. 141p.

Strickler, Dr. D. 1993. *Wayside Wildflowers of the Pacific Northwest*. Columbia Falls, MT: Flower Press. 272p.

Description

Small-flowered penstemon is a perennial forb that grows from a woody stem-base. The stems are erect, tufted, and reach 5-40 cm in height. The leaves are opposite, oval to lance shaped, with entire margins. The basal leaves have petioles while stem leaves are lacking petioles. The flowers are blue to purple, occasionally tinged pink, or white, with the petals 6-12 mm long and fused together. Flowers are borne in dense clusters along the stem and tip that appear as whorls, but are actually growing on short peduncles that arise from the leaf axil. The fruit is a 4-5 mm long caspule that contains numerous seeds (Strickler 1993, Pojar and MacKinnon 1994).

Habitat and Geographic Range

Small-flowered penstemon ranges from Alaska and northern Canada, south to Colorado, and west to California. It grows on open, rocky slopes, dry, sandy banks, grassy hillsides, and in dry meadows, and open woods at mid- to high elevations (Strickler 1993, Pojar and MacKinnon 1994).

Propagation

Seed: Seeds mature around mid-August. They require diurnal fluctuating temperatures and alternating light of 20°C (light) and 30°C (dark). Sow in March at 2 mm deep in a light soil and keep at 13-18°C for translanting outdoors in May (Prockter 1976, Redente et al. 1982).
Vegetative: Take cuttings from the node in August and place in a sandy soil in a cold frame. Protect from winter frost (Prockter 1976).

Smilacena racemosa
False Solomon's-seal

Description

False Solomon's-seal is a perennial herb that grows from stout rhizomes. The stems are erect or stiffly arched, unbranched and reaching 0.3-1 m in height. The leaves are alternate, broad, elliptical to lanceolate in shape, 7-20 cm long, with distinct parallel veins. The small, cream-colored flowers are borne in dense, egg-shaped terminal clusters. The fruit is a red, round berry, 5-7 mm across. The plant had many medicinal uses for local Native Americans (Pojar and MacKinnon 1994).

Habitat and Geographic Range

False Solomon's-seal ranges from Alaska south to California, east to Arizona, Colorado, Missouri, Mississippi, Georgia, and north to Nova Scotia. It ranges in elevation from sea level to 3050 m. It can be found in moist forests, streambanks, meadows, and shady to open forests. It grows in soils rich in organic matter (Hitchcock and Cronquist 1973, Tilford 1993, Pojar and MacKinnon 1994).

Propagation

Seed: False Solomon's-seal seed has a double dormancy and requires two years to germinate. Sow in the fall immediately after the seed has ripened. Seedlings need a relatively shaded area with consistently moist soil (Snyder 1991, Tilford 1993).
Vegetative: Plants can be propagated by rhizomes or by division in the fall or early spring. Only 7.5 cm or less of the rhizome is sufficient for propagating; plant into a moist, sterile medium such as peat moss (Snyder 1991, Tilford 1993).

References

Hitchcock, C.L., and A. Cronquist. 1973. *Flora of the Pacific Northwest: An Illustrated Manual*. Seattle, WA: University of Washington Press. 730p.

Pojar, J., and A. MacKinnon. 1994. *Plants of the Pacific Northwest Coast: Washington, Oregon, British Columbia, and Alaska*. Vancouver, BC, Canada: British Columbia Ministry of Forests and Lone Pine Publishing. 527p.

Snyder, L.C. 1991. *Native Plants for Northern Gardens*. Anderson Horticultural Library, University of Minnesota. 277p.

Tilford, G.L. 1993. *The EcoHerbalist's Fieldbook: Wildcrafting in the Mountain West*. Conner, MT: Mountain Weed Publishing. 295p.

Thalictrum fendleri
Fendler meadowrue

References

Dayton, W.A. 1960. Notes on Western Range Forbs: Equisetaceae through Fumariaceae. USDA Forest Service Handbook No. 161. Washington, D.C.: U.S. Government Printing Office. 71p.

Redente, E.F., P.R. Ogle, and N.E. Hargis. 1982. Growing Colorado Plants from Seed: A State of the Art. Vol. III: Forbs. USDI Fish and Wildlife Service.FWS/OBS-82/30. 141p.

USDA Forest Service. 1988. *Range Plant Handbook.* New York: Dover Publications, Inc. 816p.

Description

Fendler meadowrue is an erect perennial herb reaching 0.3-0.9 m in height. The leaves are alternate, ternately decompound, the lower leaves are stalked and the upper are stalkless. The leaflets are rounded to heart shaped, three lobed, and usually not longer than 1.3 cm. The flowers are small, petalless, dioecious, borne in terminal panicles, and greenish in color. The male flowers have numerous stamens with the anthers coming to an abrupt slender point at the tip. The sepals are elliptic to oblong, whitish and papery. The female flowers have eight to twelve stalkless pistils with sepals similar to the male flower. The seed is an egg-shaped, somewhat flattened, three-ribbed achene about 6 mm in length (USDA 1988).

Habitat and Geographic Range

Fendler meadowrue ranges from Montana and Idaho west to California, south to the mountains of western Texas, and into northern Mexico. It is found growing in shade in rich, moist loam soils along with aspens and shrubs. It can also be found growing in open areas and occasionally in ponderosa pine, Engelmann spruce, and other conifer forests up to 3050 m (Dayton 1960, USDA 1988).

Propagation

Seed: Store seed in airtight bottles at 2°C. It must be placed under running water for four hours to remove a brownish substance that delays germination. Stratify in a dilute solution of gibberelic acid on double layers of filter paper. Fluctuating diurnal temperatures of 22°C (8 hours with light) and 17°C (16 hours) are necessary (Redente et al. 1982).

Vicia americana
American vetch

Description
American vetch is a smooth or hairy perennial with trailing or climbing stems up to 95 cm in length. The leaves are pinnately compound and divided into eight to eighteen oval to linear leaflets with rounded to truncate or emarginate tips. The leaves contain tendrils at the ends and the stipules are sharply toothed. There are three to nine bluish-purple flowers, 15-28 mm long, arranged in loose racemes that are shorter than the leaves. The pod is strongly compressed. American vetch is highly browsed by livestock and wildlife. The seed and foliage are eaten by birds and rodents. It is a good species for soil stabilization and for revegetation of coal-mined lands and critical stabilization sites (Hermann 1966, Coladonato 1993).

Habitat and Geographic Range
American vetch is found from Alaska to Quebec, south to southern Virginia, and west across the Great Plains to California, Oregon, and Washington. It is found in moist to dry areas such as meadows, thickets, swampy woods, grassy valleys, and foothills. It grows on sandy, clayey soils and deep porous loams. American vetch is shade tolerant and occurs in all stages of succession (Hermann 1966, Coladonato 1993).

Propagation
Seed: American vetch begins growth in early spring to early summer. It flowers from May to August with seed maturing approximately one month later. Clean the seed by mechanically flailing and clipping and store at 20°C. There are no stratification or scarification requirements. Seed should be at least one year old; plant in the spring or fall in a moist, clayey soil. Keep the temperature at a constant 20°C with eight hours of light (Redente et al. 1982, Coladonato 1993).
Seeds per kilogram: ~54,010-91,080 (Redente et al. 1982)
Vegetative: American vetch is able to reproduce by creeping rhizomes (Coladonato 1993).

References
Coladonato, M. 1993. *Vicia americana. In*: Fischer, William C. (comp.) The Fire Effects Information System [Monograph Online]. Missoula, MT: USDA Forest Service, Intermountain Fire Sciences Laboratory. http://www.fs.fed.us/database/feis/plants/Forb/VICAME. Accessed February 6, 1997.

Hermann, F.J. 1966. Notes on Western Range Forbs: Cruciferae through Compositae. USDA Forest Service Handbook No. 293. Washington, D.C.: U.S. Government Printing Office. 365p.

Redente, E.F., P.R. Ogle, and N.E. Hargis. 1982. Growing Colorado Plants from Seed: A State of the Art. Vol. III: Forbs. USDI Fish and Wildlife Service. FWS/OBS-82/30. 141p.

Wyethia amplexicaulis
Mules-ear

References

Hermann, F.J. 1966. Notes on Western Range Forbs: Cruciferae through Compositae. USDA Forest Service Handbook No.293. Washington, D.C.: U.S. Government Printing Office. 365p.

Matthews, R.F. 1993. *Wyethia amplexicaulis. In*: Fischer, William C. (comp.) The Fire Effects Information System [Monograph Online]. Missoula, MT: USDA Forest Service, Intermountain Fire Sciences Laboratory. http://www.fs.fed.us/database/feis/plants/Forb/WYEAMP. Accessed February 7, 1997.

Description

Mules-ear is a stout perennial forb. The leathery, glossy basal leaves are oblong to lanceolate in shape with entire or toothed margins, and grow 40 cm long and 15 cm wide. The stem leaves are smaller, sessile, and usually clasping the stem. Flower heads are large and several in number, and the involucral bracts are smooth. The fruit is an achene. The taproot can be up to 22 cm in circumference and reach depths of over 180 cm. Lateral roots grow 90-120 cm from the main root. Mules-ear is very aromatic. Young basal leaves, flower heads, and seeds are grazed by livestock. Dense stands provide good cover for birds and small mammals. Its well developed root system make it a good species for soil stabilization. It may also be useful for revegetating mine sites (Hermann 1966, Matthews 1993).

Habitat and Geographic Range

Mules-ear ranges from southwestern Montana to eastern Washington and Oregon, and south to southeastern Wyoming, Colorado, northern Utah, and northeastern Nevada. It is found above the sagebrush zone, 1360-3300 m. It grows in meadows, open woods, moist draws, and open grasslands. It grows on a variety of soils from sandy loams to clay loams, but does best in heavy clay soils. It requires 25-45 cm of precipitation annually (Hermann 1966, Matthews 1993).

Propagation

Seed: Mules-ear flowers from April through July and seed matures during July and August. Seed can be collected by hand-pulling the heads and cleaned by drying and fanning. Seed remains viable for up to five years. Although seed will germinate without stratification, it should undergo a cool, moist stratification at 2-5°C for four weeks to enhance germination. Diurnal temperatures of 10°C (for 16 hours) and 25°C (for 8 hours) also promote maximum germination (Plummer et al. 1968, Young and Evans 1979, Matthews 1993).

Seeds per kilogram: ~54,290 (Plummer et al. 1968).

Vegetative: Mules-ear sprouts from underground rootstalks or from the root crown following damage to aboveground portions of the plant (Matthews 1993).

Plummer, A.P., D.R. Christensen, and S.B. Monsen. 1968. Restoring Big Game Range in Utah. Utah Div. Fish Game Publ. 68-3. 183p.

Young, J.A., and R.A. Evans. 1979. Arrowleaf balsamroot and Mules-ear seed germination. *Journal of Range Management* 32(1):71-74.

Xerophyllum tenax
Common beargrass

Description

Common beargrass is a rigid, tufted, evergreen, herbaceous perennial that grows 1-2 m in height. The wiry leaves are grasslike with rough edges, and are 30-90 cm long. The flower clusters occur at the top of the stalk, and have a broad base tapering to a blunt point. Hundreds of white flowers are crowded together on slender, elongated white pedicles. The flowers have a heavy, slightly unpleasant fragrance and bloom once every five to seven years. The fruit is a small, three-lobed capsule containing several 4-mm-long seeds. The flower stalks are eaten by deer, elk, and other big game animals. Common beargrass has high biomass production and is good for long-term revegetation and erosion control. Coastal and Southwest tribes use the leaves in basketweaving (Dayton 1960, Crane 1990).

Habitat and Geographic Range

Common beargrass can be found in all soil types ranging from basaltic lava to serpentine to granite and quartzite. It prefers a well-drained soil and grows on xeric to subxeric and submesic to mesic sites. It is moderately shade tolerant and can be found in the understory of cool subalpine forests or in the open on high ridges and slopes and in mountain meadows. It is usually found between 900 and 2450 m in elevation but grows from sea level up to 2700 m. It ranges from central California north to British Columbia east to southwestern Alberta, and south through the Rocky Mountains into Idaho, Montana, and northwestern Wyoming (Dayton 1960, Crane 1990).

Propagation

Seed: Common beargrass blooms from July to August and fruiting occurs during August and September (Crane 1990). Seed can be collected by hand in the fall and sown directly (Anderson 1996) or stored dry at subfreezing temperatures (Smart and Minore 1977). For spring sowing, presoak the seed in distilled water for 24 hours, sow on moist vermiculite with a light covering of more vermiculite, and cold stratify for 16 weeks at 3°C. Set the flats in a growth chamber at 18°C (for 12 hours)

References

Albright, M. 1996. Greenhouse Manager. USDI National Park Service. Olympic National Park, Port Angeles, WA. Personal communication.

Anderson, J. 1996. Sevenoaks Native Nursery, Corvallis, OR. Personal communication.

Crane, M.F. 1990. *Xerophyllum tenax. In*: Fischer, William C. (comp.) The Fire Effects Information System [Monograph Online]. Missoula, MT: USDA Forest Service, Intermountain Fire Sciences Laboratory. http://www.fs.fed.us/database/feis/plants/Forb/XERTEN. Accessed September 11, 1996.

during the day and 13°C (for 12 hours) at night (Smart and Minore 1977). Seed also germinates well if it is soaked and sown on a peat:vermiculite:perlite:pumice (2:2:2:1) medium, covered with 1-2 cm of perlite, and wetted down. Flats can be covered with an air- and water-permeable fabric and placed outside in a sheltered location until spring. Uncover the flats in the spring prior to germination and move into a warm location in the greenhouse once germination begins. It is important to keep a layer of dry perlite on the surface of the flats since beargrass seedlings are susceptible to damping-off. Seedlings can be transplanted into deep pots with a well-drained medium and covered with a thin layer of perlite or pumice around the root crown to prevent crown rot. Care should be taken when transplanting not to injure the young roots that sprout adventitiously at the base of the crown (Albright 1996).

Seeds per kilogram: ~1,829,805 (Crane 1990).

Vegetative: Vegetative reproduction occurs by offshoots of the rhizome. These offshoots can be divided in the spring or fall and replanted (Crane 1990, Time Life, Inc. 1996).

Dayton, W.A. 1960. Notes on Western Range Forbs: Equisetaceae through Fumariaceae. USDA Forest Service Handbook No. 161. Washington, D.C.: U.S. Government Printing Office, 71p.

Smart, A.W., and D. Minore. 1977. Germination of beargrass (*Xerophyllum tenax* [Pursh.] Nutt.). *The Plant Propagator* 23(3):13-15.

Time Life, Inc. Electronic Encyclopedia. Virtual Garden. [Monograph Online]. http://pathfinder.com/@@yJuDF4F68wAAQAJ6/cgi-bin/VG/vg. Accessed June 25, 1996.

Grasses and Sedges

Agropyron spicatum
(Pseudoroegneria spicata)
Bluebunch wheatgrass

References

Archibald, C., and S. Feigner. 1995. USDA Forest Service, J. Herbert Stone Nursery, Central Point, OR. Personal communication.

Hitchcock, A.S. 1971. *Manual of the Grasses of the United States*. Vol. I and II. New York: Dover Publications, Inc. 1051p.

Miller, R.F., J.M. Seufert, and M.R. Haferkamp. 1986. The ecology and management of bluebunch wheatgrass (*Agropyron spicatum*): A review. Agric. Exp. Sta., Oregon State University Station Bulletin No. 669. 39p.

Description

Bluebunch wheatgrass is a green or glaucous perennial with tufted, erect culms, often in large bunches, growing 60-100 cm tall. The sheaths are glabrous and the blades are flat to loosely involute, 1-4 mm wide, glabrous beneath, and pubscent above. The spike is slender and 8-15 cm long. The rachis is scaberulous on the angles, and the internodes are 1-2 cm long. The spikelets are distant, not as long as the internodes, six- to eight-flowered, and the rachills joints are scaberulous and 1.5-2 mm long. The glumes are narrow, obtuse to acute, about four-nerved, glabrous or scabrous on the nerves, and usually half as long as the spikelet. The lemmas are about 1 cm long, the awn is 1-2 cm long and strongly divergent, and the palea is obtuse and about as long as the lemma (Hitchcock 1971). Bluebunch wheatgrass provides palatable forage for livestock and wildlife.

Habitat and Geographic Range

Bluebunch wheatgrass is drought resistant and is found on plains and dry slopes, and in canyons and dry open woods of the western United States. It is found from northern Michigan to Alaska, south to western South Dakota, New Mexico, and California, and throughout the Great Basin. It grows well on deep, well-drained, loamy soils but is adapted to coarser textured soils. It is the climax vegetation of the Pacific Northwest and Intermountain states, forming up to 60% of the vegetative cover in many areas (Hitchcock 1971, Miller et al. 1986).

Propagation

Seed: Flowering and seed production of bluebunch wheatgrass can be erratic. Seeds are best if collected fresh in the fall (Archibald and Feigner 1995). Collect seed heads with a combine, binder, hand stripper, or sickle. Process with a hammermill to remove the awns and produce clean seeds (Wheeler 1950). It is very important to remove the awns or they will slow sowing. No stratification is required and fall sowing is recommended due to an increase in yield over spring sown seed. Sow seeds to a depth of 0.6-1 cm and cover with a sawdust mulch. A minimum of 100-160 seeds per square meter is recommended (Archibald and Feigner 1995). Moderate fertilization may increase vigor, yield, and seed production (Miller et al. 1986).

Seeds per kilogram: ~132,275-154,320 (Archibald and Feigner 1995)

Wheeler, W.A. 1950. *Forage and Pasture Crops: A Handbook of Information about the Grasses and Legumes Grown for Forage in the United States*. The Field Seed Institute of North America. New York: D. Van Nostrand Company. 752p.

Bromus carinatus (B. marginatus)
California brome

Description

California brome is an annual or mostly biennial erect grass, with culms 5-10 dm tall and a deep fibrous root system. The sheaths are scabrous to sparsely pilose. The blades are flat, 20-30 cm long, the lower blades are shorter, 3-10 mm wide, and scabrous or sparsely pilose. The panicle is 15-30 cm long with spreading or drooping branches. The spikelets are 2-3 cm long, six- to ten-flowered, the florets scarcely overlapping in anthesis exposing the long rachilla joints. The glumes are acuminate; the first glume is 6-9 mm long and the second is 10-15 mm long. The lemmas are minutely appressed-pubescent to glabrous, 2-2.5 mm wide as folded, and 10-20 mm long. The awn is 7-15 mm long and the palea is acuminate, nearly as long as the lemma, and the teeth are short awned (Hitchcock 1971). It is one of the best forage grasses due to its resistance to drought and grazing and is browsed by all classes of livestock (USDA 1988).

Habitat and Geographic Range

California brome grows in moist woods, meadows, sagebrush hills, foothills, and mountains (Hitchcock 1971). It is found in western North America from Alaska, British Columbia, and Alberta south to California, New Mexico, and into Baja California. It ranges in elevation from 600 to 3050 m (USDA 1988).

Propagation

Seed: Seeds mature in June to early July at lower elevations and July to August at higher elevations. Collect seeds by hand stripping from plants, after checking for smut. Clean in a hammermill and store in a cool, dry place (Link 1993). No stratification is required (Jebb 1995). Seed can be direct seeded (Link 1993). Optimum germination occurs at temperatures from 15 to 25°C. Smut can be a problem and can be treated with Vitavax®(Archibald and Feigner 1995).
Seeds per kilogram: ~88,180-180,780 (Link 1993, Archibald and Feigner 1995)

References

Archibald, C., and S. Feigner. 1995. USDA Forest Service, J. Herbert Stone Nursery, Central Point, OR. Personal communication.

Hitchcock, A.S. 1971. *Manual of the Grasses of the United States*. Vol. I and II. New York: Dover Publications, Inc. 1051p.

Jebb, T. 1995. Horticulturalist, USDI Bureau of Land Management, C.A. Sprague Seed Orchard, Merlin, OR. Personal communication.

Link, E. (ed.) 1993. Native Plant Propagation Techniques for National Parks: Interim Guide. East Lansing, MI: Rose Lake Plant Materials Center. 240p.

USDA Forest Service. 1988. *Range Plant Handbook*. New York: Dover Publications, Inc. 816p.

Bromus vulgaris
Columbia brome

Description
Columbia brome is a perennial grass with slender culms, 80-120 cm tall, with pubescent nodes. The sheaths are pilose, the ligule is 3-5 mm long, and the blades are pilose and up to 12 mm wide. The panicle is 10-15 cm long; the branches are slender and drooping. The spikelets are narrow and about 2.5 cm long. The glumes are narrow; the first is acute, one-nerved, 5-8 mm long, and the second is broader, longer, obtuse to acute, and three-nerved. The lemmas are 8-10 mm long, sparsely pubescent over the back and near the margin, or nearly glabrous. The awn is 6-8 mm long (Hitchcock 1971). Coumbia brome is good forage for livestock and is grazed by elk (Walsh 1994).

Habitat and Geographic Range
This species is found in rocky woods and shady ravines, usually below 1200 m. It ranges from British Columbia to the Sierra Nevadas in California, east to Wyoming (Hitchcock 1971).

Propagation
Seed: Seed matures from July to August; collect by hand, clean in a hammermill, and store in a cool, dry place (Link 1993). No stratification is required. Sow in 3-cubic-inch containers in a peat:vermiculite (1:1) medium (Jebb 1995) or plant to a depth of 0.6-1 cm and cover with a sawdust mulch. Smut can be a problem and can be treated with Vitavax® applied at 4 oz/100 lbs of seed (Archibald and Feigner 1995).
Seeds per kilogram: ~175,485-263,450 (Link 1993)
Vegetative: Columbia brome sprouts from perennating buds at the base of the culms (Walsh 1994).

References
Archibald, C., and S. Feigner. 1995. USDA Forest Service, J. Herbert Stone Nursery, Central Point, OR. Personal communication.

Hitchcock, A.S. 1971. *Manual of the Grasses of the United States.* Vol. I and II. New York: Dover Publications, Inc. 1051p.

Jebb, T. 1995. Horticulturalist, USDI Bureau of Land Management, C.A. Sprague Seed Orchard, Merlin, OR. Personal communication.

Link, E. (ed.) 1993. Native Plant Propagation Techniques for National Parks: Interim Guide. East Lansing, MI: Rose Lake Plant Materials Center. 240p.

Walsh, R.A. 1994. *Bromus vulgaris. In*: Fischer, William C. (comp.) The Fire Effects Information System [Monograph Online]. Missoula, MT: USDA Forest Service, Intermountain Fire Sciences Laboratory. http://www.fs.fed.us/database/feis/plants/Graminoid/BROVUL. Accessed March 26, 1997.

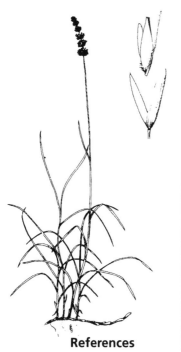

Calamagrostis rubescens
Pinegrass

Description

Pinegrass is a perennial with slender, tufted culms that reach 60-100 cm tall and have creeping rhizomes. The sheaths are smooth but pubescent on the collar. The blades are erect, 2-4 mm wide, flat to somewhat involute, and scabrous. The panicle is narrow, spikelike or somewhat loose or interrupted, pale to purple in color, and 7-15 cm long. The glumes are 4-5 mm long, narrow, and acuminate. The lemma is pale, thin, 3.5-4.5 mm long, smooth, the nerves are obscure, the awn is near the base, geniculate, exerted from the side of the glumes, 1-2 mm long above the bend, with the callus hairs scant and about one-third as long as the lemma. The rachilla is 1 mm long with the sparse hairs extending to 2 mm. Pinegrass is forage for deer, elk, pronghorn, black bear, and cattle. It may be a useful species for rehabilitating disturbed sites (Hitchcock 1971, Snyder 1991).

Habitat and Geographic Range

Pinegrass ranges from Manitoba west to British Columbia and south through Washington, Oregon, Idaho, Montana, Wyoming, Utah, Colorado, and into southern California. It ranges in elevation from sea level to 3750 m. It is moderately shade tolerant to very intolerant and is found in open pine woods, prairies, under forest canopies, and on banks. It grows best in sandy loams, loams, and clay loams although it will tolerate gravelly sand and acid soils. It is considered a late seral to climax species (Hitchcock 1971, Snyder 1991).

Propagation

Seed: Pinegrass flowers from June to September. There are no stratification or scarification requirements. Diurnal fluctuating temperatures of 20°C (for 16 hours) and 30°C (for 8 hours) are needed for germination. In addition, adequate light and moisture are required (Fulbright et al. 1982, Snyder 1991).
Vegetative: Pinegrass reproduces mainly from rhizomes (Snyder 1991).

References

Fulbright, T.E., E.F. Redente, and N.E. Hargis. 1982. Growing Colorado Plants from Seed: A State of the Art. Vol II: Grasses and grasslike plants. USDI Fish and Wildlife Service.FSW/OBS-82/29. 113p.

Hitchcock, A.S. 1971. *Manual of the Grasses of the United States*. Vol. I and II. New York: Dover Publications, Inc. 1051p.

Snyder, S.A. 1991. *Calamagrostis rubescens. In*: Fischer, William C. (comp.) The Fire Effects Information System [Monograph Online]. Missoula, MT: USDA Forest Service, Intermountain Fire Sciences Laboratory. http://www.fs.fed.us/database/feis/plants/Graminoid/CALRUB. Accessed February 6, 1997.

Carex obnupta
Slough sedge

Description

Slough sedge is a somewhat clustered, erect, evergreen perennial growing 60-150 cm tall. The culms are rather stout, triangular in cross section, and bear three to seven pendulate floral spikes 1.5-5 cm long. The leaves are mostly basal, flat, deep green, 2-6 mm wide and up to 3 dm long. The margins are rolled under with an obvious channel running along the axis. The upper one to three spikes of the inflorescence are male and often curving. The lower two to four spikes are cylindrical and female, or may have male flowers at the top. They are found on short erect stalks or are occasionally stalkless. The perigynia are shiny, yellowish green or brown in color, elliptical, plump, leathery, thick-walled and two-ribbed but nerveless on both sides, with a very short beak. The scales are purplish-black, with sharp tips, and are longer than the perigynia. Native people still use slough sedge to make baskets (Peck 1961, Weinmann et al. 1984, Pojar and MacKinnon 1994).

Habitat and Geographic Range

Slough sedge is found in shallow fresh marshes and swamps. It can also be found in bogs and wet forest openings and on streambanks and coastal shores. It is common at low elevations and ranges from British Columbia south to California west of the Cascade Mountains (Peck 1961, Pojar and MacKinnon 1994).

Propagation

Seed: Seed can be collected by hand in June and July and requires little to no cleaning other than to separate the seed from the seedhead. Store dry in a refrigerator. Sow in the fall and keep wet outdoors for natural stratification or stratify wet in a refrigerator and sow in February in potting soil and place into a hoop house. Germination usually occurs within a month (Buis 1996).

References

Buis, S. 1996. Owner, Sound Native Plants, Olympia, WA. Personal communication.

Peck, M.E. 1961. *A Manual of the Higher Plants of Oregon.* Portland, OR: Binfords & Mort Publishers. 936p.

Pojar, J., and A. MacKinnon. 1994. *Plants of the Pacific Northwest Coast: Washington, Oregon, British Columbia, and Alaska.* Vancouver, BC, Canada: British Columbia Ministry of Forests and Lone Pine Publishing. 527p.

Weinmann, F., M. Boulé, K. Brunner, J. Malek, and V. Yoshino. 1984. Wetland Plants of the Pacific Northwest. US Army Corps of Engineers, Seattle District. 85p.

Carex rostrata
Beaked sedge

Description

Beaked sedge is a large grasslike perennial that grows in tufts from short root stocks. The culms are stout, 40-120 cm high, sharp angled only above, and usually much shorter than the leaves. The leaves are nodulose, flat, 5-12 mm wide, with soft and spongy sheaths. There are four to ten leaves per stem. The bracts are leaflike, very long and usually surpassing the inflorescence. Beaked sedge has two types of spikes: a terminal, relatively small staminate, and a lower, axillary, larger pistillate. The seed is often three angled and enclosed in a saclike body called the perigynium. The perigynia are ovoid, 4-6 mm long, inflated, strongly nerved, straw-colored or brownish, abruptly narrowed to cylindric, have a strongly bidentate beak, and are a third to a fourth as long as the body. Beaked sedge stands are important feeding and breeding grounds for geese and waterfowl and are often grazed by elk, moose, and reindeer (Peck 1961, USDA 1988, Cope 1992).

Habitat and Geographic Range

Beaked sedge is very common in wet meadows and marshes, and on edges of lakes, ponds, streams, and other riparian areas. It grows in a variey of soils. It is found in cool and semi-arid climates at elevations of 750-3350 m. It ranges from Greenland to Alaska, southward to Arizona, and east to Kansas and Delaware (Cope 1992).

Propagation

Seed: The shoots of beaked sedge emerge during July and August. The flora primordia develop in August and September and the shoot flowers during June and July of the following year and then usually dies in late August or September (Cope 1992). Harvest the achenes during August and September by clipping the inflorescence from the plant. Press the perigynia between the thumb and index finger to test for the presence of an achene inside. Dry the inflorescences by spreading them out over a fine screen in a warm, dry, well-ventilated area. Clean seed by hand stripping and using a sieve or airscreen cleaner to remove chaff. Purity

References

Buis, S. 1996. Owner, Sound Native Plants, Olympia, WA. Personal communication.

Cope, A.B. 1992. *Carex rostrata*. *In*: Fischer, William C. (comp.) The Fire Effects Information System [Monograph Online]. Missoula, MT: USDA Forest Service, Intermountain Fire Sciences Laboratory. http://www.fs.fed.us/database/feis/plants/Graminoid/CARROT. Accessed September 10, 1996.

can be improved with a seed blower. Seed can be stored in sealed containers at room temperature with a moisture content of 6-8% for up to 17 months (Hurd and Shaw 1992). Sow in the fall and keep wet outdoors for natural stratification or stratify wet in a refrigerator and sow in February in potting soil and place into a hoop house. Germination usually occurs within a month (Buis 1996)

Vegetative: Beaked sedge reproduces by rhizomes that range from 1 cm to 2.5 m in length. A long rhizome first emerges and produces a shoot, then short rhizomes develop to produce a tuft of many shoots. These give the plant a matted and tufted growth pattern. When shoots first develop, they do not have any roots. Beaked sedge can also reproduce by stolons (Cope 1992).

Hurd, E.G., and N.L. Shaw. 1992. Seed technology for *Carex* and *Juncus* species of the Intermountain region. *In:* Proceedings, Intermountain Forest Nursery Association. Park City, UT; August 12-16, 1991. USDA Forest Service, Gen. Tech. Rep. RM-211. 119p.

Peck, M.E. 1961. *A Manual of the Higher Plants of Oregon.* Portland, OR: Binfords & Mort Publishers. 936p.

USDA Forest Service. 1988. *Range Plant Handbook.* New York: Dover Publications, Inc. 816p.

Danthonia californica
California oatgrass

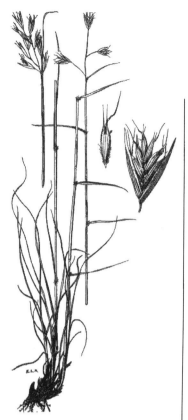

References

Guerrant, E.O. Jr., and A. Raven. 1995. Seed germination and storability studies of 69 plant taxa native to the Willamette Valley Wet Prairie. Portland, OR: The Berry Botanic Garden. 66p.

Hitchcock, A.S. 1971. *Manual of the Grasses of the United States*. Vol. I and II. New York: Dover Publications, Inc. 1051p.

Jebb, T. 1995. Horticulturalist, USDI Bureau of Land Management, C.A. Sprague Seed Orchard, Merlin, OR. Personal communication.

Description

California oatgrass is a perennial with culms 30-100 cm tall, glabrous, and tending to disarticulate at the nodes. The sheaths are glabrous and pilose at the throat. The blades are 10-20 cm long, glabrous, and flat, or as most of the innovations, involute. The panicle bears two to five spikelets; the pedicels are slender, spreading or somewhat reflexed, flexuous, 1-2 cm long, with a prominent pulvinus at the base of each. The glumes are 15-20 mm long; the lemmas, excluding the awns, are 8-10 mm long, glabrous, pilose on the lower part of the margin and on the callus, and with long aristate teeth. The terminal segment of the awn is 5-10 mm long, and the palea is subacute, usually extending beyond the base of the awn (Hitchcock 1971). This species is moderate to good forage for livestock (USDA 1988).

Habitat and Geographic Range

California oatgrass grows in meadows and open woods (Hitchcock 1971). It ranges from British Columbia to Montana, Colorado, and California, and can grow up to 3050 m in elevation (USDA 1988).

Propagation

Seed: California oatgrass seeds require a one- to three-day soak in running water and three months cold stratification at 1-5°C (Jebb 1995). Guerrant and Raven (1995) also report good germination by cold stratification at 5°C for six weeks followed by a diurnally fluctuating warm stratification of 20°C (for 16 hours) and 10°C (for 8 hours) for another six weeks. Plant seeds in 3-cubic-inch containers in a peat:vermiculite (1:1) medium and fertilize once a week with a low nitrogen fertilizer. Seed does not have a good germination rate (Jebb 1995).
Seeds per kilogram: ~201,770 (Guerrant and Raven 1995)

Vegetative: California oatgrass can be propagated by division. Collect plants during the dormant season, or leave them in a lathhouse until dormant. Bring them into a greenhouse in January and divide into plants each with a single root system. Pot into containers and keep moist in a greenhouse with an air temperature of 18-21°C. After two weeks, plants can be moved back to a lathhouse (Link 1993).

Link, E. (ed.) 1993. Native Plant Propagation Techniques for National Parks: Interim Guide. East Lansing, MI: Rose Lake Plant Materials Center. 240p.

USDA Forest Service. 1988. *Range Plant Handbook*. New York: Dover Publications, Inc. 816p.

Danthonia intermedia
Timber oatgrass

Description

Timber oatgrass is a perennial with culms 10-50 cm tall. The sheaths are glabrous with long hairs on the throat. The blades are subinvolute, or flat as are those on the culm, glabrous or sparsely pilose. The panicle is purplish, narrow, with few flowers, 2-5 cm long, with appressed branches bearing a single spikelet. The glumes are about 15 mm long, the lemmas are 7-8 mm long, appressed-pilose along the margin below and on the callus, the summit is scaberulous, and the teeth are acuminate and aristate tipped. The terminal segment of the awn is 5-8 mm long, and the palea is narrowed above and notched at the apex (Hitchcock 1971). Timber oatgrass provides forage for livestock and wildlife, especially in the spring because it greens up before many other plants begin growth (USDA 1988).

Habitat and Geographic Range

Timber oatgrass is found in meadows and bogs in northern and alpine regions (Hitchcock 1971). It ranges from Newfoundland and Quebec to Alaska, south to northern Michigan, New Mexico, and California from lowlands to subalpine areas (Link 1993, Taylor and Douglas 1995).

Propagation

Seed: Flowering occurs from June to August and seed matures during August and September. Clean seeds with an air screen cleaner. Seed stored dry at 1-7°C will remain viable for up to five years (Link 1993). No stratification is required. Plant in the fall in small containers in a peat:vermiculite (1:1) medium and set outside. After germination, plants require nitrogen fertilizer. Transplant in the spring (Archibald and Feigner 1995).

Seeds per kilogram: ~1,172,840 (Link 1993).

References

Archibald, C., and S. Feigner. 1995. USDA Forest Service, J. Herbert Stone Nursery, Central Point, OR. Personal communication.

Hitchcock, A.S. 1971. *Manual of the Grasses of the United States.* Vol. I and II. New York: Dover Publications, Inc. 1051p.

Link, E. (ed.) 1993. Native Plant Propagation Techniques for National Parks: Interim Guide. East Lansing, MI: Rose Lake Plant Materials Center. 240p.

Taylor, R.J. and G.W. Douglas. 1995. *Mountain Plants of the Pacific Northwest: A Field Guide to Washington, Western British Columbia, and Southeastern Alaska.* Missoula, MT: Mountain Press Publishing Company. 437p.

USDA Forest Service. 1988. *Range Plant Handbook.* New York: Dover Publications, Inc. 816p.

Deschampsia atropurpurea
Mountain hairgrass

Description

Mountain hairgrass is a perennial with loosely tufted, erect culms that are purplish at the base and grow 40-80 cm in height. The blades are flat, soft, ascending or appressed, 5-10 cm long, 4-6 mm wide, and acute or abruptly acuminate. The panicle is loose, open, 5-10 cm long, and the few capillary drooping branches are naked below. The spikelets are purplish and broad. The glumes are broad and about 5 mm long. The second glume is three-nerved and exceeds the florets. The lemmas are scabrous, about 2.5 mm long; the callus hairs are one-third to half as long; the awn of the first is straight and included; the second is geniculate and exerted (Hitchcock 1971).

Habitat and Geographic Range

Mountain hairgrass ranges from Newfoundland south to the White Mountains of New Hampshire, west to Alaska, and south to California and Colorado. It grows in high-elevation meadows, woods, alpine heath, stream banks, and open subalpine forests (Hitchcock 1971, Pojar and MacKinnon 1994).

Propagation

Seed: Collect seed in August and September, hand clean, and store at 7°C. Stratify by placing between moist paper at 5°C for 18 weeks. It is also reported that no stratification is required. Sow in a peat:vermiculite (1:1) mixture in containers and grow for three months. Use a low-nitrogen fertilizer (9-45-15) once a week (Jebb 1995, Kaye 1997).

References

Hitchcock, A.S. 1971. *Manual of the Grasses of the United States.* Vol. I and II. New York: Dover Publications, Inc. 1051p.

Jebb, T. 1995. Horticulturalist, USDI Bureau of Land Management, C.A. Sprague Seed Orchard, Merlin, OR. Personal communication.

Kaye, T.N. 1997. Seed dormancy in high elevation plants: implications for ecology and restoration. pp. 115-20 *In*: Conservation and Management of Native Plants and Fungi. T.N. Kaye, A. Liston, R.M. Love, D.L. Louma, R.J. Meinke, and M.V. Wilson (eds.). Corvallis, OR: Native Plant Society of Oregon.

Pojar, J., and A. MacKinnon. 1994. *Plants of the Pacific Northwest Coast: Washington, Oregon, British Columbia, and Alaska.* Vancouver, BC, Canada: British Columbia Ministry of Forests and Lone Pine Publishing. 527p.

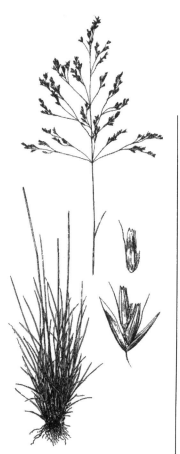

Deschampsia caespitosa
Tufted hairgrass

Description

Tufted hairgrass is a perennial with culms in dense tufts, erect, leafy at the base, and 60-120 cm tall. The sheaths are smooth, the blades are 1.5-4 mm wide, often elongate, firm, flat or folded, and scabrous above. The panicle is loose, open, nodding, 10-25 cm long, and the capillary has scabrous branches and branchlets which bear spikelets toward the ends. The spikelets are 4-5 mm long, pale or purplish tinged, the florets are distent, and the rachilla internode is half the length of the lower floret. The glumes are one-nerved or the second glume is obscurely three-nerved, acute, and about as long as the florets. The lemmas are smooth and the callus hairs short. The awn is found near the base and can be straight and included in the glumes or weakly geniculate. The awn is twice as long as the spikelets (Hitchcock 1971). It is grazed by all classes of livestock and is occasionally cut for hay (USDA 1988).

Habitat and Geographic Range

Tufted hairgrass grows in nearly pure stands in well-watered areas, and mixed stands in meadows and open spaces. It occurs mainly in the spruce-fir belt and above timberline, and can be found in the mountains in all of the western states (USDA 1988).

Propagation

Seed: Fields of tufted hairgrass should be swathed in mid- to late June and allowed to cure in the field for seven to ten days. Clean the seed by using an air screen machine to remove weed seeds, empty hulls, straw, and chaff. Delint using a standard debearder or a huller-scarifier. Store seeds under cool temperatures and low humidity (Darris et al. 1995). No stratification is required (Archibald and Feigner 1995). However, seed that was kept at 20°C (for 16 hours) and 10°C (for 8 hours) for six weeks, then transfered to a cooler environment of 15°C (for 16 hours) and 5°C (for 8 hours) had a slightly higher germination rate (Guerrant and Raven 1995). Sow in either fall or spring. Recommended row spacing is 45-75 cm and, based on 200-400 pure live seeds per linear meter and a row spacing of 60 cm, the recommended

References

Archibald, C. and S. Feigner. 1995. USDA Forest Service, J.Herbert Stone Nursery, Central Point, OR. Personal communication.

Darris, D., S. Lambert, and W. Young III. 1995. Seed production of tufted hairgrass. Portland, OR: USDA Natural Conservation Service. Plant Materials Techn. Note 16. 4p.

seeding rate is 1-2 kg/ha. Sow 0.6 cm deep. Periodic sprinkler irrigation during May through August helps establish spring-seeded stands. Apply fertilizer after the seedlings are established (Darris et al. 1995).

Seeds per kilogram: ~3,658,510 (Guerrant and Raven 1995)

Vegetative: Tufted hairgrass can be propagated by division in the fall or spring (Greenlee 1992).

Greenlee, J. 1992. *The Encyclopedia of Ornamental Grasses: How to Grow and Use over 250 Beautiful and Versatile Plants.* New York: Rodale Press. 186p.

Guerrant, E.O. Jr. and A. Raven. 1995. Seed germination and storability studies of 69 plant taxa native to the Willamette Valley Wet Prairie. Portland, OR: The Berry Botanic Garden. 66p.

Hitchcock, A.S. 1971. *Manual of the Grasses of the United States.* Vol. I and II. New York: Dover Publications, Inc. 1051p.

USDA Forest Service. 1988. *Range Plant Handbook.* New York: Dover Publications, Inc. 816p.

Elymus elymoides (Sitanion hystrix)
Squirreltail

References

Archibald, C., and S. Feigner. 1995. USDA Forest Service, J. Herbert Stone Nursery, Central Point, OR. Personal communication.

Hitchcock, A.S. 1971. *Manual of the Grasses of the United States.* Vol. I and II. New York: Dover Publications, Inc. 1051p.

Jebb, T. 1995. Horticulturalist, USDI Bureau of Land Management, C.A. Sprague Seed Orchard, Merlin, OR. Personal communication.

Link, E. (ed.) 1993. Native Plant Propagation Techniques for National Parks: Interim Guide. East Lansing, MI: Rose Lake Plant Materials Center. 240p.

Stubbendieck, J., S.L. Hatch, and C.H. Butterfield. 1992. *North American Range Plants.* Lincoln, NE: University of Nebraska Press. 493p.

Young, J.A., and R.A. Evans. 1977. Squirreltail seed germination. *Journal of Range Management* 30(1): 33-36.

Description

Squirreltail is a perennial that grows 10-50 cm in height and has stiff, erect to spreading culms. The foliage varies from glabrous or puberulent to softly and densely white pubescent. The blades are flat to involute, 5-20 cm long, 1-3 mm wide, and stiffly ascending to spreading. The spike is erect, 2-7 cm long, short exserted or partly included, and the glumes are very narrow, one- to two-nerved with the nerves extending into scabrous awns, sometimes bifid to the middle, or bear a bristle or awn along one margin. The lemmas are convex, smooth or scabrous to appressed pubescent, and the awns of both the glumes and lemmas are widely spreading and 2-10 cm long (Hitchcock 1971). Squirreltail is valuable to livestock and wildlife in late summer and early fall after the inflorescences have broken and fallen (Stubbendieck et al. 1992).

Habitat and Geographic Range

Squirreltail can be found on dry hills, plains, open woods, and rocky slopes. It ranges from British Columbia and Alberta to southern California, and east to South Dakota (Hitchcock 1971).

Propagation

Seed: Seed is produced in the first year (Archibald and Feigner 1995) and matures from June to September depending on latitude and altitude (Link 1993). Seeds are difficult to collect and clean due to the long awns (Archibald and Feigner 1995) so it is best to collect by hand, de-awn with a brush machine, and airscreen (Link 1993). Seeds with high viability germinate rapidly with no stratification. Plant seeds into a moist seedbed to a depth of 0.6-1 cm, cover with a sawdust mulch, and water every ten days (Archibald and Feigner 1995, Jebb 1995). Optimum temperature regimes are a constant 15°C, or alternating 10°C (for 16 hours, night) and 15°C (for 8 hours, day), or alternating 10°C (for 16 hours, night) and 20°C (for 8 hours, day) (Young and Evans 1977).

Seeds per kilogram: ~220,455-396,825 (Archibald and Feigner 1995)

Elymus glaucus
Blue wildrye

Description

Blue wildrye is a perennial with culms in loose to dense tufts, often bent at the base, erect, 60-120 cm tall, leafy, and without rhizomes. The sheaths are smooth or scabrous. The blades are flat, usually lax, 8-15 mm wide, scabrous on both surfaces, and sometimes narrow and subinvolute. The spike is long exserted, erect to somewhat nodding, usually dense, and 5-20 cm long. The glumes are lanceolate at the base, 8-15 mm long, with two to five strong scabrous nerves, and acuminate or awn-pointed. The lemmas are awned, the awn one to two times as long as the body, and erect to spreading (Hitchcock 1971). Blue wildrye can provide fair to good forage for big game and livestock early in the growing season. As a pioneer species, it is recommended for erosion control on steep, eroded slopes, roadsides, or fire damaged sites (Darris et al. 1996).

Habitat and Geographic Range

Blue wildrye grows in prairies, open woods, thickets, and moist or dry hillsides from sea level to 3000 m. It ranges from Alaska to southern California, east to Ontario, Michigan, and Iowa, and south to Colorado (Hitchcock 1971).

Propagation

Seed: Harvest seed by hand and clean by gently de-awning with a brush machine or air screen machine. It can be stored for up to five years if kept under low temperatures and low humidity (Link 1993). No pretreatment is required for adequate germination. Sow directly in the fall to a depth of 0.6-1 cm at a density of 100-160 seeds per square meter (Archibald and Feigner 1995) and cover with a sawdust mulch. Spring sow from March through early May. The seedbed should be moist, fine-textured, very firm, and weed free. Germination is rapid, usually within six to ten days. A nitrogen starter fertilizer is recommended, and a complete fertilizer when seedlings are well established (Darris et al. 1996).
Seeds per kilogram: ~198,410-262,345 (Archibald and Feigner 1995)
Vegetative: Blue wildrye can also be propagated by division (Greenlee 1992).

References

Archibald, C., and S. Feigner. 1995. USDA Forest Service, J. Herbert Stone Nursery, Central Point, OR. Personal communication.

Darris, D.C., S.M. Lambert, and W.C. Young, III. 1996. Seed production of blue wildrye. Portland, OR: USDA Natural Resources Conservation Service. Plant Materials Tech. Note No. 17. 5p.

Greenlee, J. 1992. *The Encyclopedia of Ornamental Grasses: How to Grow and Use over 250 Beautiful and Versatile Plants.* New York: Rodale Press. 186p.

Hitchcock, A.S. 1971. *Manual of the Grasses of the United States.* Vol. I and II. New York: Dover Publications, Inc. 1051p.

Link, E. (ed.) 1993. Native Plant Propagation Techniques for National Parks: Interim Guide. East Lansing, MI: Rose Lake Plant Materials Center. 240p.

Festuca viridula
Green-leaf fescue

Description

Green-leaf fescue is a perennial with loosely tufted, erect culms growing 50-100 cm tall. The blades are soft, erect, those of the culm are flat or loosely involute, and those of the innovations are slender and involute. The panicle is open, 10-15 cm long, the branches mostly in pairs, ascending or spreading, slender, somewhat remote, and naked below. The spikelets are three- to six-flowered, the glumes are lanceolate, somewhat unequal, and 5-7 mm long. The lemmas are membranaceous, acute or cuspidate, glabrous, and 6-8 mm long (Hitchcock 1971). It is foraged by all classes of livestock throughout the grazing season (USDA 1988).

Habitat and Geographic Range

Green-leaf fescue is found in mountain meadows and on open slopes at elevations of 1000-2000 m. It ranges from British Columbia to Alberta, south to central California and Idaho, and is found in Colorado (Hitchcock 1971).

Propagation

Seed: Seed matures from June to September, depending on latitude and elevation, and can be collected by hand or with a sickle. Clean seed with an airscreen and store cool and dry. Seeds should undergo a cold, moist stratification for 16 weeks prior to greenhouse planting. Seeds can be also be sown directly in the fall (Link 1993). **Seeds per kilogram:** ~1,653,440 (Link 1993)

References

Hitchcock, A.S. 1971. *Manual of the Grasses of the United States.* Vol. I and II. New York: Dover Publications, Inc. 1051p.

Link, E. (ed.) 1993. Native Plant Propagation Techniques for National Parks: Interim Guide. East Lansing, MI: Rose Lake Plant Materials Center. 240p.

USDA Forest Service. 1988. *Range Plant Handbook.* New York: Dover Publications, Inc. 816p.

Hordeum brachyantherum
Meadow barley

Description
Meadow barley is a perennial, tufted grass that grows 20-70 cm tall. The lower sheaths are thin, often shredded with soft hairs; blades are 3-8 mm wide, and 8-10 cm long. The floret of the central spikelet is usually 7-10 mm long and 1.5 mm wide, the awn is about 1 cm long, while the glumes are slightly shorter. The glumes of the lateral spikelets are usually unequal; the floret varies from well developed and staminate to much reduced and empty, the awn is 1-1.5 cm long. There are three spikelets per node with the central spikelet being stalkless (Hitchcock 1971, Pojar and MacKinnon 1994).

Habitat and Geographic Range
Meadow barley ranges from Alaska south to California, east to New Mexico, and north into Montana. It is also found in Maine, New Hampshire, Ohio, Indiana, and Mississippi. It can be found in meadows, bottom lands, and salt marshes, and on ocean beaches and grassy slopes from sea level up to 3000 m in elevation (Hitchcock 1971, Pojar and MacKinnon 1994).

Propagation
Seed: Collect seed in late summer and store in a paper bag in a cold, dry environment. Meadow barley seed germinates easily and with a high percentage (~96%). Sow fresh seed in the fall at a depth of two to three times its height in a sand:pumice:peat (1:1:1) medium and place in a cold frame. Seed can also be cold stratified at 5°C for six weeks, then transferred to temperatures of 10°C (8 hours a day in the dark) and 20°C (16 hours of light) for an additional six weeks. It is also reported that fluctuating temperatures of 20°C (for 16 hours a day) and 30°C (for 8 hours) are required for germination (Fulbright et al. 1982, Guerrant and Raven 1995).
Seeds per kilogram: ~148,660 (Guerrant and Raven 1995)

References
Fulbright, T.E., E.F. Redente, and N.E. Hargis. 1982. Growing Colorado Plants from Seed: A State of the Art. Vol II: Grasses and Grasslike Plants. USDI Fish and Wildlife Service. FSW/OBS-82/29. 113p.

Guerrant, E.O. Jr., and A. Raven. 1995. Seed germination and storability studies of 69 plant taxa native to the Willamette Valley Wet Prairie. Portland, OR: The Berry Botanic Garden. 66p.

Hitchcock, A.S. 1971. *Manual of the Grasses of the United States*. Vol. I and II. New York: Dover Publications, Inc. 1051p.

Pojar, J., and A. MacKinnon. 1994. *Plants of the Pacific Northwest Coast: Washington, Oregon, British Columbia, and Alaska*. Vancouver, BC, Canada: British Columbia Ministry of Forests and Lone Pine Publishing. 527p.

Koeleria cristata
(K. macrantha)
Junegrass

Description

Junegrass, also known as Koeler's grass, is a loosely tufted perennial that grows 30-60 cm tall with erect culms. The sheaths are pubescent, the blades are flat or involute, glabrous, lower blades are pubescent, and are 1-3 mm wide. The panicle is erect, spikelike, dense, often lobed, interrupted, or sometimes branched below, 4-15 cm long, and tapering at the summit. The spikelets are 4-5 mm long, sometimes short-awned, and the rachilla joints are very short (Hitchcock 1971). It is considered good forage for all classes of livestock and is grazed early in the season (USDA 1988).

Habitat and Geographic Range

Junegrass ranges from Ontario to British Columbia, south to California, Mexico, Louisiana, Missouri, and Delaware. It can be found in open woods, prairies, and sandy soils (Hitchcock 1971).

Propagation

Seed: Seed from Junegrass is produced during the second year. No stratification is required. Plant in the fall to a depth of 0.5-1.0 cm and cover with a sawdust mulch. This species develops a rust; treat with a fungicide (Archibald and Feigner 1995).
Seeds per kilogram: ~5,000,000 (Archibald and Feigner 1995)

References

Archibald, C., and S. Feigner. 1995. USDA Forest Service, J. Herbert Stone Nursery, Central Point, OR. Personal communication.

Hitchcock, A.S. 1971. *Manual of the Grasses of the United States.* Vol. I and II. New York: Dover Publications, Inc. 1051p.

USDA Forest Service. 1988. *Range Plant Handbook.* New York: Dover Publications, Inc. 816p.

Melica harfordii
Harford's melic

Description
Harford's melic is a perennial, 60-120 cm tall, with tufted culms that are often decumbent below. The sheaths are scabrous to villous with scabrous, firm blades that are 1-4 mm wide and flat to subinvolute in shape. The panicle is narrow, 10-15 cm long with appressed branches. The spikelets are short pediceled and 1-1.5 cm long, and the glumes are 7-9 mm long, and obtuse. The lemmas are faintly seven-nerved, hispidulous below, and pilose on the lower part of the margin. The apex is emarginate, mucronate, or with an awn that is less than 2 mm long (Hitchcock 1971).

Habitat and Geographic Range
Harford's melic ranges from British Columbia to the Cascade Mountains in Oregon, south to Monterey County, California. It grows in open dry woods and slopes (Hitchcock 1971).

Propagation
Seed: Seed is produced during the first growing season and cleans easily off the plant. No stratification is required. Planted in the fall 0.6-1 cm deep and cover with a sawdust mulch. No fungal problems are associated with this species (Archibald and Feigner 1995).

References
Archibald, C., and S. Feigner. 1995. USDA Forest Service, J. Herbert Stone Nursery, Central Point, OR. Personal communication.

Hitchcock, A.S. 1971. *Manual of the Grasses of the United States*. Vol. I and II. New York: Dover Publications, Inc. 1051p.

Phleum alpinum
Alpine timothy

References

Fulbright, T.E., E.F. Redente, and N.E. Hargis. 1982. Growing Colorado Plants from Seed: A State of the Art. Vol II: Grasses and Grasslike Plants. USDI Fish and Wildlife Service. FWS/OBS-82/29. 113p.

Hitchcock, A.S. 1971. *Manual of the Grasses of the United States.* Vol. I and II. New York: Dover Publications, Inc. 1051p.

Kaye, T.N. 1997. Seed dormancy in high elevation plants: implications for ecology and restoration. pp. 115-20 *In*: Conservation and Management of Native Plants and Fungi. T.N.Kaye, A. Liston, R.M. Love, D.L. Louma, R.J. Meinke, and M.V. Wilson (eds.). Corvallis, OR: Native Plant Society of Oregon.

Stubbendieck, J., S.L. Hatch, and C.H. Butterfield. 1992. *North American Range Plants.* Lincoln, NE: University of Nebraska Press. 493p.

USDA Forest Service. 1988. *Range Plant Handbook.* New York: Dover Publications, Inc. 816p.

Description

Alpine timothy is a perennial with erect, glabrous culms 15-60 cm tall that stem from a decumbent, somewhat creeping, densely tufted base. The blades are flat, with the margins and lower surface scabrous, and grow up to 15 cm long and 3-8 mm wide. The panicle is bristly, dense, purplish in color, and ellipsoid or short-cylindric in shape. The glumes are subequal, 3-5 mm long, hispid-ciliate on the keel, with 2 mm long awns. The spikelets are one-flowered, elliptic, and somewhat flattened. The lemma is glabrous and shorter than the glumes, and the apex is slightly erose. Alpine timothy is an excellent browse for livestock as well as big game (Hitchcock 1971, Stubbendieck et al. 1992).

Habitat and Geographic Range

Alpine timothy ranges from Alaska east to Alberta and south to New Mexico in the west and from Labrador south to New Hampshire and into northern Michigan in the east. It can be found above 1250 m in mountain meadows, bogs, and other wet places. It grows in deep, poor to well-drained soils and is fairly shade tolerant (Hitchcock 1971, Fulbright et al. 1982, USDA 1988, Stubbendieck et al. 1992).

Propagation

Seed: Collect seed in August and September, hand clean, and store at 7°C. It can be sown in either fall or spring with excellent germination. Temperature for germination should remain at a constant 20°C (Fulbright et al. 1982, Kaye 1997). After-ripen spring-sown seed for nine months at 7°C.

Poa scabrella
Pine bluegrass

Description

Pine bluegrass grows 50-100 cm tall with erect culms that are usually scabrous below the panicle. The blade is mostly basal, 1-2 mm wide, lax, and scabrous, the ligule is 3-5 mm long, and the sheath is scaberulous. The panicle is narrow, 5-12 cm long, and contracted. The spikelets are 6-10 mm long, the glumes are 3 mm long and scabrous, and the lemmas are 4-5 mm long, and crisp-puberulent on the back toward the base. Pine bluegrass is an important browse plant for livestock (Hitchcock 1971, USDA 1988).

Habitat and Geographic Range

Pine bluegrass ranges from western Montana and Colorado, west to Washington, and south to California and Baja California. It can be found at low to mid-elevations in meadows, open woods, rocky areas, and hills (Hitchcock 1971).

Propagation

Seed: Seed matures from June to September. Fulbright et al. (1982) reports that seed should be prechilled for two weeks. Jebb (1995) recommends using potassium nitrate and light for germination. Keep the temperature at a constant 29°C. Sow seed in small containers in a peat:vermiculite (1:1) medium and grow for three months (Fulbright et al. 1982, Jebb 1995).
Seeds per kilogram: ~2,204,585-2,865,965 (Archibald and Feigner 1995)

References

Archibald, C., and S. Feigner. 1995. USDA Forest Service, J. Herbert Stone Nursery, Central Point, OR. Personal communication.

Fulbright, T.E., E.F. Redente, and N.E. Hargis. 1982. Growing Colorado Plants from Seed: A State of the Art. Vol II: Grasses and Grasslike Plants. USDI Fish and Wildlife Service. FWS/OBS-82/29. 113p.

Hitchcock, A.S. 1971. *Manual of the Grasses of the United States.* Vol. I and II. New York: Dover Publications, Inc. 1051p.

Jebb, T. 1995. Horticulturalist, USDI Bureau of Land Management, C.A. Sprague Seed Orchard, Merlin, OR. Personal communication.

USDA Forest Service. 1988. *Range Plant Handbook.* New York: Dover Publications, Inc. 816p.

Poa secunda (sandbergii)
Sandberg bluegrass

References

Bradley, A.F. 1986. *Poa secunda. In*: Fischer, William C. (comp.) The Fire Effects Information System [Monograph Online]. Missoula, MT: USDA Forest Service, Intermountain Fire Sciences Laboratory. http://www.fs.fed.us/database/feis/plants/Graminoid/POASEC. Accessed March 13, 1997.

Fulbright, T.E., E.F. Redente, and N.E. Hargis. 1982. Growing Colorado Plants from Seed: A State of the Art. Vol II: Grasses and grasslike plants. USDI Fish and Wildlife Service. FWS/OBS-82/29. 113p.

Hitchcock, A.S. 1971. *Manual of the Grasses of the United States*. Vol. I and II. New York: Dover Publications, Inc. 1051p.

Stubbendieck, J., S.L. Hatch, and C.H. Butterfield. 1992. *North American Range Plants*. Lincoln, NE: University of Nebraska Press. 493p.

USDA Forest Service. 1988. *Range Plant Handbook*. New York: Dover Publications, Inc. 816p.

Description
Sandberg bluegrass grows 10-60 cm tall with erect culms that grow from a dense, extensive tuft of short basal foliage. The blades are short, soft, flat, folded or involute, with the ligule acute and prominent. The sheaths are rounded, glabrous, and persistent. The panicle is 2-10 cm long and narrow, yellowish-green to purple in color, not densely flowered, and the branches are short and appressed, becoming somewhat spreading during anthesis. The spikelets are two to five flowered, 4-6 mm long, terate, and acute. The glumes are unequal, papery; the first glume is one- to three- nerved and 2.2-5 mm long, and the second is three-nerved and 3-4 mm long. The lemmas are minutely scabrous, and crisp-pubescent near the base. There are no awns. It is good forage for cattle and fair for other livestock and deer (Hitchcock 1971, Stubbendieck 1992).

Habitat and Geographic Range
Sandberg bluegrass ranges from Alaska and the Yukon Territory south to North Dakota and New Mexico west to southern California. It can be found at elevations of 300-3660 m. It grows in dry woods, and open timber areas, and on plains, grassy slopes, ridge tops, and rocky slopes. It can grow in a variety of soils ranging from deep sandy or silt loams to shallow rocky soils. It is one of the most drought-resistant grasses of the bluegrass species (Hitchcock 1971, USDA 1988, Stubbendieck 1992).

Propagation
Seed: Seed is the only means of regeneration for Sandberg bluegrass. It matures in early summer and can be harvested with a small grain combine. Seed can be planted in the fall with no pretreatment required. Fluctuating diurnal temperatures of 12°C (for 16 hours a day) and 17°C (for 8 hours) are required for germination. Do not plant deeper than 3 cm in a clay loam or sandy soil (Fulbright et al. 1982, Bradley 1986).
Seeds per kilogram: ~1,984,125-2,039,245 (Fulbright et al. 1982)

Stipa lemmonii
(Achnatherum lemmonii)
Lemmon needlegrass

Description

Lemmon needlegrass is a tufted perennial with culms 30-80 cm tall that are scaberulous and usually puberulent below the nodes. The ligule is 1-3 mm long. The blades are 10-20 cm long, 1-2 mm wide, flat or involute, and the blades of the innovations are very narrow. The panicle is 5-12 cm long, narrow, and pale or purplish in color. The glumes are 8-10 mm long, broad and firm, abruptly acuminate, with the first being five-nerved and the second three-nerved. The lemma is 6-7 mm long, pale or light brown, with a rather blunt callus, fusiform body, 1.2 mm wide, and is villous with appressed hairs. The awn is 20-35 mm long, twice-geniculate, and appressed-pubescent to the second bend. It is a valuable forage plant and grazed mainly when young (Hitchcock 1971).

Habitat and Geographic Range

Lemmon needlegrass ranges from British Columbia south to Idaho and California. It is found on dry open ground and in open woods (Hitchcock 1971).

Propagation

Seed: Collect seed in early June by hand stripping. Store in a paper bag and keep refrigerated through the summer. It is important that the seed be cleaned well and de-awned. Cold stratify in potassium nitrate and gibberellic acid and plant in 3-cubic-inch containers in a peat:vermiculite (1:1) medium. Apply a low-nitrogen fertilizer once a week (Jebb 1995).

References

Hitchcock, A.S. 1971. *Manual of the Grasses of the United States.* Vol. I and II. New York: Dover Publications, Inc. 1051p.

Jebb, T. 1995. Horticulturalist, USDI Bureau of Land Management, C.A. Sprague Seed Orchard, Merlin, OR. Personal communication.

Stipa occidentalis
(Achnatherum occidentalis)
Western needlegrass

Description
Western needlegrass is a tufted perennial that grows up to 12 dm tall and has a fibrous root system. The ligules are 0.5-1.0 mm long. The flowers are arranged in a very narrow panicle with the spikelets being one flowered. The lemmas are a soft pubescent and are awned from the tip. The awns are 1.5-5 cm long, twisted and bent twice. The plant flowers from May to August and has good palatability to cattle, sheep, and deer (Helliwell 1987).

Habitat and Geographic Range
Western needlegrass prefers well-drained soils in clearings, open forests, and meadows, and near old roads. It ranges from the Yukon/British Columbia to southern California, east to the Dakotas, and south to New Mexico from the upper foothills to higher elevations (Helliwell 1987, USDA 1988, Link 1993).

Propagation
Seed: Seed matures in late August to early September and can be hand harvested. Clean with a brush machine, de-awn with a mantle brush, then airscreen with medium air flow. Store seed under cool, dry conditions. Sow shallowly into small containers in the spring with five to eight seeds per container, which should be left outside for one week to allow for natural chilling, then move inside a greenhouse. Direct field sowing in the spring produces poor results due to weed competition. Survival is enhanced by sowing seed in fall in media amended with peat and slow-release fertilizer (Link 1993).
Seeds per kilogram: ~685,625 (Link 1993)

References
Helliwell, R. 1987. Forest Plants of the Warm Springs Indian Reservation. Warm Springs, OR: Confederated Tribes of the Warm Springs. 177p.

Link, E. (ed.) 1993. Native Plant Propagation Techniques for National Parks: Interim Guide. East Lansing, MI: Rose Lake Plant Materials Center. 240p.

USDA Forest Service. 1988. *Range Plant Handbook*. New York: Dover Publications, Inc. 816p.

Shrubs

Acer circinatum
Vine maple

Description

Vine maple is a long-lived, deciduous shrub or small tree growing up to 15 m in height. The trunk is crooked and the branches grow erect or creep along the ground before turning up. Stems are pale green to reddish brown. The simple, opposite leaves appear nearly circular with seven to nine sharply toothed lobes. Leaves are reddish when young and turn bright green, then orange and scarlet in late summer or fall. Flowers are found in loose, hanging clusters and are reddish-purple or white in color. The fruits are winged, rose-colored double samaras at angles of 140-180°. It is not a commercially important species, but is sometimes used as an ornamental or for firewood. Vine maple is eaten by deer, elk, cattle and sheep. The seeds, buds, and flowers are a source of food for many birds and small mammals (Elias 1980, Uchytil 1989, Haeussler et al. 1990, King County 1994).

Habitat and Geographic Range

Vine maple is very shade tolerant and grows in the understory of coniferous forests. It can be found from sea level to 1400 m in elevation and ranges from the Cascade Mountains to the coast and from southwestern British Columbia to northern California. It prefers bottomlands and rich, moist, well-drained soils and can be found growing along streambanks and alluvial terraces, in forest openings and clearcuts, and on talus slopes and the lower portions of open slopes (Elias 1980, Uchytil 1989, Haeussler 1990).

Propagation

Seed: Vine maple is a very poor seed producer and relies primarily on vegetative means for natural reproduction (Uchytil 1989). It does, however, begin to produce seed at an early age, probably before age ten. Collect the seeds by hand or by shaking the tree and collecting the samaras on a tarp. Seed collection should be done in September and October as the samaras begin to dry. Seeds require a warm, moist stratification at 20-30°C for thirty to sixty days, followed by a cold stratification at 3°C for 90-180 days. Plant in trays or beds and cover

References

Elias, T.S. 1980. *The Complete Trees of North America Field Guide and Natural History.* New York: Van Nostrand Reinhold Company. 948p.

Haeussler, S., D. Coates, and J. Mather. 1990. Autecology of common plants in British Columbia: A literature review. British Columbia Ministry of Forests. FRDA Report-158. 272p.

King County Department of Public Works, Surface Water Management Division. 1994. Northwest Native Plants, Identification and Propagation for Revegetation and Restoration Projects. King County, WA. 68p.

Olson, D.F. Jr., and W.J. Gabriel. 1974. *Acer* L. Maple. pp. 187-94 *In:* Schopmeyer, C.S. (tech. coord.) 1974. *Seeds of the Woody Plants in the United States.* Agric. Handbook 450. Washington, DC: USDA Forest Service. 883p.

Uchytil, R. J. 1989. *Acer circinatum. In*: Fischer, William C. (comp.) The Fire Effects Information System [Monograph Online]. Missoula, MT: USDA Forest Service, Intermountain Fire Sciences Laboratory. http://www.fs.fed.us/ database/ feis/plants/ Tree/ACECIR. Accessed June 25, 1996.

Wilson, M.G. 1996. Restoration Ecologist, Portland, OR. Personal communication.

with leaf mulch. They will germinate the following spring and can be transplanted after one year (Olson and Gabriel 1974, Haeussler 1990). West of the Cascade Mountains, direct seed into deep flats covered with mulch and place in a cold frame or bury at ground level and keep through the winter. The seed will germinate the following spring and seedlings can be transplanted one year later (Wilson 1996).

Seeds per kilogram: ~7,690-12,195 (Olson and Gabriel 1974)

Vegetative: Plants sprout from the root crown following top kill from logging or burning. Layering occurs infrequently, but as plants mature some stems become too long and massive to remain erect and thus lay prostrate and root where the stem touches the ground. Seedlings growing around an abundant supply of vine maples may be dug up and potted immediately (Uchytil 1989, King County 1994).

Acer glabrum
Douglas maple (Rocky Mountain maple)

Description

Douglas maple is a shade-tolerant, deciduous small tree or shrub that grows up to 9 m. The trunk is short with smooth, reddish-brown to gray bark. The leaves are simple, opposite, palmate shaped with three to five toothed lobes. The male and female flowers usually occur on separate trees and can be found in hanging clusters at the end of the branchlets. Fruit are paired, winged samaras usually at an angle of 45° or less. Rodents and birds eat the seeds while whitetail and mule deer browse the leaves and twigs. Due to the small size of the tree, the wood is not commercially important, but is excellent for campfires and wood stoves (Elias 1980).

Habitat and Geographic Range

Douglas maple is found on wet sites at elevations between 1200 and 1800 m. It grows in sheltered canyons and ravines, on moist slopes, and along streams. It is commonly found on silty to sandy or gravelly and rocky well-drained soils. Due to its flexible stems, it can withstand heavy snow pack and often codominates avalanche chutes. Its range is from southern Alaska to New Mexico (Elias 1980, Uchytil 1989).

Propagation

Seed: Douglas maple begins to produce seed as early as ten years of age. Seeds reach maturity from August through early October. Hand picking of the samaras is the best method of collection. Clean by hand rubbing or hammermilling of the wings and blowing off chaff. Dry the seeds to 10-15% moisture content and store at 2-5°C in sealed containers. Warm stratification for 180 days followed by cold stratification for 180 days gives a 25% germination rate in container seeding in a greenhouse. Seed can also be sown by direct field planting in the fall (Olson and Gabriel 1974).

Seeds per kilogram: ~17,235-44,755 (Olson and Gabriel 1974)

Vegetative: Douglas maple sprouts easily from root crowns following a disturbance (Olson and Gabriel 1974).

References

Elias, T.S. 1980. *The Complete Trees of North America Field Guide and Natural History.* New York: Van Nostrand Reinhold Company. 948p.

Olson, D.F. Jr., and W.J. Gabriel. 1974. *Acer* L. Maple. pp. 187-94 *In*: Schopmeyer, C.S. (tech. coord.) 1974. *Seeds of the Woody Plants in the United States.* Agric. Handbook 450. Washington, DC: USDA Forest Service. 883p.

Uchytil, R.J. 1989. *Acer glabrum. In*: Fischer, William C. (comp.) The Fire Effects Information System [Monograph Online]. Missoula, MT: USDA Forest Service, Intermountain Fire Sciences Laboratory. http://www.fs.fed.us/database/feis/plants/Tree/ACEGLA. Accessed June 25, 1996.

Amelanchier alnifolia
Saskatoon serviceberry

Description

Saskatoon serviceberry is a deciduous shrub ranging from single stems 20 cm tall to treelike clumps up to 6 m tall. The alternate leaves are 2.5-5 cm long, oval shaped, serrate at the margin, and have distinct veins from the midrib to the teeth. The flowers are white with five long petals, and the small fruit is a dark blue pome. Deer eat the leaves and twigs while the fruits are eaten by both birds and mammals (Elias 1980).

Habitat and Geographic Range

Saskatoon serviceberry can be found either singly or in a thicket. Its range extends from Alaska south to California, New Mexico, northeast to the Dakotas, Michigan, and into western Ontario. It grows from near sea level to over 2750 m. It prefers sunny or partially shaded areas with moist to dry soils (Elias 1980, USDA 1988).

Propagation

Seed: A seed crop is produced every three to five years and should be collected in late summer. The best method of collection is to knock the fruit onto a canvas or directly into hoppers. Extract seeds by macerating in water and washing over screens and clean by drying and rubbing through the screen, then running through a fanning mill. Store seeds dry in a sealed container at 5°C for no more than five years (Brinkman 1974). Captan® may be added, since serviceberry has a tendency for fungal molds (Macdonald 1986). Plant seeds in the winter or stratify with a moist chill at 1-6°C for four to six months and plant in the spring. Sow in a sandy soil to a depth of 0.5 cm and keep mulched until germination, which usually takes place the second spring (Vories 1981).
Seeds per kilogram: ~80,025-250,885 (Brinkman 1974)

References

Brinkman, K.A. 1974. *Amelanchier* Med. Serviceberry. pp. 212-15 *In*: Schopmeyer, C.S. (tech. coord.) 1974. *Seeds of the Woody Plants in the United States.* Agric. Handbook 450. Washington, DC: USDA Forest Service. 883p.

Elias, T.S. 1980. *The Complete Trees of North America Field Guide and Natural History.* New York: Van Nostrand Reinhold Company. 948p.

Macdonald, B. 1986. *Practical Woody Plant Propagation for Nursery Growers.* Portland, OR: Timber Press. 669p.

Vegetative: Saskatoon serviceberry can be vegetatively propagated by both root cuttings and division. Take root cuttings during the dormant season, optimally December to February. Take one-year-old, fleshy roots the diameter of a pencil, from as close to the crown of the plant as possible. Cut roots to 5 cm in length with a horizontal cut at the proximal end and a slanted cut at the distal end. Treat cuttings with a fungicide before sticking vertically in rows, 5 cm apart, with the proximal end at soil level, and covered with 1.5 cm of perlite. Division is best if done in early spring. Remove suckers by cutting with a sharp spade. Wash off excess soil. Cut back the stem and trim the root system. Plant the division in pots, beds, or open ground. Adequate irrigation is required to prevent the roots from drying out (Macdonald 1986).

USDA Forest Service. 1988. *Range Plant Handbook*. New York: Dover Publications, Inc. 816p.

Vories, K.C. 1980. Growing Colorado Plants from Seed: A State of the Art. Vol. 1: Shrubs. Ogden, UT: USDA Forest Service Intermountain Forest and Range Experiment Station Gen. Tech. Rep. INT-103. 80p.

Arctostaphylos nevadensis
Pinemat manzanita

References

Carlson, J.R., and W.C. Sharp. 1975. Germination of high elevation manzanitas. *Tree Planters' Notes* 26(3):10-11; 25.

Helliwell, R. 1987. Forest Plants of the Warm Springs Indian Reservation. Warm Springs, OR: Confederated Tribes of the Warm Springs. 177p.

Hitchcock, C.L., and A. Cronquist. 1973. *Flora of the Pacific Northwest: An Illustrated Manual.* Seattle, WA: University of Washington Press. 730p.

Peck, M.E. 1961. *A Manual of the Higher Plants of Oregon.* Portland, OR: Binfords & Mort Publishers. 936p.

Pojar, J., and A. MacKinnon. 1994. *Plants of the Pacific Northwest Coast: Washington, Oregon, British Columbia, and Alaska.* Vancouver, BC, Canada: British Columbia Ministry of Forests and Lone Pine Publishing. 527p.

Trindle, J.D.C. 1995. Evaluating acid scarification effects on dormant *Arctostaphylos nevadensis* seeds. International Plant Propagators Society 45:312-14.

Van Dersal, W.R. 1938. Native Woody Plants of the United States: Their Erosion-Control and Wildlife Values. USDA Misc. Pub. 303. 362p.

Description

Pinemat manzanita is a low-growing, creeping evergreen shrub with branches reaching 50-120 cm long and 10-30 cm high. It has an exfoliating reddish bark. The leaves are alternate, obovate to oblanceolate, 16-25 mm long, and pointed at the tip. The flowers are white, urn-shaped, and are borne in small clusters at the ends of the branches. The fruit is a brownish-red, berrylike drupe 5-8 mm in diameter. The berries are eaten by bear, grouse, turkeys, and other wildlife (Peck 1961, Helliwell 1987).

Habitat and Geographic Range

Pinemat manzanita ranges from California north to the Cascade Mountains of Oregon and southern Washington and east to the Blue Mountains of Oregon. It occurs widely at higher elevations (Hitchcock and Cronquist 1973, Pojar and MacKinnon 1994).

Propagation

Seed: Collect berries from late summer to early fall, clean, dry, and store in the dark at 5-15°C. Seeds need both a scarification and stratification treatment. Studies show that soaking the seed for 25 minutes in concentrated sulfuric acid is enough to erode the micropyle plug without damaging the embryonic tissue. Sow in a peat:sand (1:1) medium. Even after pretreatments, germination rates are very low (Carlson and Sharp 1975, Trindle 1995).

Vegetative: Pinemat manzanita can be propagated by layering (Van Dersal 1938).

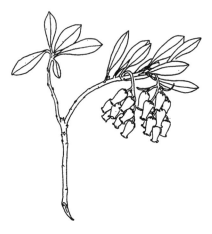

Arctostaphylos patula
Greenleaf manzanita

Description
Greenleaf manzanita is an erect, evergreen shrub that grows up to 2 m in height. The bark is smooth and reddish-brown and shreds with age, exposing the whitish-green wood underneath. The many-branched limbs are crooked, stout, and rigid. The leaves are alternate, ovate to elliptical in shape, 3-5 cm long, entire, and glossy, light green. The pinkish flowers are urn shaped, and grow in terminal clusters. The fruit is a reddish-brown to black berrylike drupe with four to ten stony seeds. Deer use greenleaf manzanita for cover (Helliwell 1987).

Habitat and Geographic Range
Greenleaf manzanita grows on dry, well-drained sandy loam to silty loam soils. It can be found from 950 to 3050 m in elevation from Mt. Hood, Oregon, south into southern California, Nevada, and Arizona (Randall et al. 1994, Helliwell 1987, Zimmerman 1991).

Propagation
Seed: Greenleaf manzanita produces heavy seed crops annually. Flowering is from late March to June depending on location and seeds reach maturity from July to September. It is best to collect during late fruit development by hand or by picking fruit off the ground. Clean the fruit by macerating and separating the nutlets by floatation or blowing (Berg 1974). Seeds require hot water scarification (Jebb 1995) followed by cold stratification at 4°C for 90 days in moist sand (Berg 1974). Sulfuric acid has not been found to be effective (Jebb 1995). Seeds germinate in the laboratory at alternating temperatures of 30°C during the day, and 20°C at night. Seeds can also be sown in early summer in coarse soil and kept mulched over winter (Berg 1974).
Seeds per kilogram: ~39,685 (Berg 1974)

References
Berg, A.R. 1974. *Arctostaphylos* Adans. Manzanita. pp. 228-31 *In*: Schopmeyer, C.S. (tech. coord.) 1974. *Seeds of the Woody Plants in the United States*. Agric. Handbook 450. Washington, DC: USDA Forest Service. 883p.

Helliwell, R. 1987. Forest Plants of the Warm Springs Indian Reservation. Warm Springs, OR: Confederated Tribes of the Warm Springs. 177p.

Jebb, T. 1995. Horticulturalist, USDI Bureau of Land Management, C.A. Sprague Seed Orchard, Merlin, OR. Personal communication.

Randall, W.R., R.F. Keniston, D.N. Bever, and E.C. Jensen. 1994. *Manual of Oregon Trees and Shrubs*. Corvallis, OR: Oregon State University Bookstores. 305p.

Zimmerman, M.L. 1991. *Arctostaphylos patula. In*: Fischer, William C. (comp.) The Fire Effects Information System [Monograph Online]. Missoula, MT: USDA Forest Service, Intermountain Fire Sciences Laboratory. http://www.fs.fed.us/database/feis/plants/Shrub/ARCPAT. Accessed March 26, 1997.

Vegetative: Greenleaf manzanita is easiest to propagate from stem cuttings of the current year's growth. Take cuttings to a length of five to six nodes with a slice cut at the bottom node. Remove the lower leaves and dip the cuttings in a rooting hormone. Stick the cuttings in a rooting medium of peat:perlite (1:1) and keep them moist (Jebb 1995). Greenleaf manzanita regenerates naturally through layering or sprouting of dormant buds within the root burl (Zimmerman 1991).

Arctostaphylos uva-ursi
Kinnickinnick

Description

Kinnickinnick, also called bearberry, is a low-growing shrub, 5-20 cm in height, with woody stems. The branches grow up to 75 cm in length and form a mat that can cover a large area of ground. The leaves are oblong to ovate, 15-20 mm long, 10 mm wide, dark leathery green, and arranged opposite on the reddish purple woody stems. The pinkish white flowers are 4-5 mm long, urn shaped, and droop in a short terminal raceme. The fruit is a round drupe, 7-12 mm in diameter, dry in texture, and contains five nutlets (Govt. of Saskatchewan 1989). Kinnickinnick is browsed by deer, elk, bighorn sheep, and moose. The fruits are eaten by birds, deer, elk, small mammals, and bears. This species is good for erosion control on slopes (MacKenzie 1989).

Habitat and Geographic Range

Kinnickinnick grows on a variety of soils although it is most commonly found on well-drained sandy and gravelly soils (Randall et al. 1994). It prefers sun or partial shade, and is usually found under pine (Helliwell 1987). It can be found from sea level to 2500 m in elevation. It ranges from northern California north to Alaska and across Canada and the northern United States to New England and Newfoundland. In the west, it extends south in the Rocky Mountains to New Mexico, and in the east, it extends south along the Atlantic coast to New Jersey and in the Appalachian Mountains to Virginia (Randall et al. 1994, Helliwell 1987).

Propagation

Seed: The fruit turns bright red or pink upon ripening from June to August. Collect by hand from the plants or off the ground. Clean by maceration and separate the nutlets by flotation or blowing. Seeds require an acid scarification treatment which can be accomplished by immersion in sulfuric acid for three to six hours. After scarification, seeds should undergo a warm stratification for 60 days followed by a cold stratification for 60 days. Sow in early summer and mulch over winter (Berg 1974).

Seeds per kilogram: ~59,080-83,555 (Berg 1974)

References

Berg, A.R. 1974. *Arctostaphylos* Adans. Manzanita. pp. 228-31 *In*: Schopmeyer, C.S. (tech. coord.) 1974. *Seeds of the Woody Plants in the United States*. Agric. Handbook 450. Washington, DC: USDA Forest Service. 883p.

Government of Saskatchewan. 1989. Guide to Forest Understory Vegetation in Saskatchewan. Canada Forestry. Technical Bulletin No. 9/1980 revised January, 1989. 106p.

Helliwell, R. 1987. Forest Plants of the Warm Springs Indian Reservation. Warm Springs, OR: Confederated Tribes of the Warm Springs. 177p.

Holloway, P. and J. Zasada. 1979. Vegetative propagation of 11 common Alaska woody plants. USDA Forest Service Pacific Northwest Forest and Range Experiment Station. Research Note PNW-334. 12p.

MacKenzie, D.S. 1989. *Arctostaphylos uva-ursi. American Nurseryman* 170(4):194.

Randall, W.R., R.F. Keniston, D.N. Bever, and E.C. Jensen. 1994. *Manual of Oregon Trees and Shrubs.* Corvallis, OR: Oregon State University Bookstores. 305p.

Vegetative: Take stem cuttings in early spring or early fall. They should consist of all available new growth plus a portion of one-year-old tissue. Dip in an IBA talc, plant immediately into a sandy medium, and place in an intermittent mist irrigation greenhouse at a temperature of 22°C. Bottom heat and mycorrhizal inoculation can help rooting success (MacKenzie 1989). Kinnikinnick can also be propagated by root cuttings taken in the fall. Bury 5-10 cm long root cuttings horizontally in peat, moisten, and place in a greenhouse (Holloway and Zasada 1979).

Arctostaphylos viscida
Whiteleaf manzanita

Description:

Whiteleaf manzanita is an erect, long-lived, evergreen shrub. It ranges from 1-4 m high, with spreading branches covering an average area of 1.5 m². Its bark is smooth, reddish-brown in color, and continually shed. The leaves are whitish-green, stiff, leathery with entire margins, and broadly ovate to elliptical in shape. The flowers are white or pinkish, urn shaped, and found in a sticky-stemmed cluster. Its fruit is a light red, flatly globose drupe containing hard-coated seeds. The laterally spreading, shallow roots usually penetrate less than 20 cm below ground. Whiteleaf manzanita fruit is a food source for various wildlife animals including black bear, coyote, brush rabbit, and dusky grouse. Black-tailed deer browse sprouts or seedlings and sometimes the older leaves in the winter. Dense stands provide good cover and nesting sites for small birds and mammals (Howard 1992, Randall et al. 1994).

Habitat and Geographic Range:

Whiteleaf manzanita occurs in California and Oregon. It is found in the foothills of the Sierra Nevada Mountains, in the north Coast Ranges, Klamath Ranges, and Siskiyou Mountains. Whiteleaf manzanita ranges in elevation from 150 to 1525 m. It is shade intolerant, seldom grows under mature conifer canopies, and is typically found on dry, sunny slopes. The plant occurs in a Mediterranean climate, with mild, wet winters and hot, dry summers. It grows in shallow, rocky, sandy soil though some populations have adapted to serpentine soil (Howard 1992).

Propagation

Seed: Whiteleaf manzanita flowers from February to April. Fruits appear in early summer and ripen in late summer or early fall. Seed is dispersed from late summer until the following spring (Howard 1992). Seed requires heat and mechanical or chemical scarification followed by an overwinter stratification. Seeds are produced annually, although production slows during drought years (Howard 1992).

Vegetative: All manzanita species can regenerate by layering (Howard 1992).

References

Howard, J.L. 1992. *Arctostaphylos viscida. In:* Fischer, William C. (comp.) The Fire Effects Information System [Monograph Online]. Missoula, MT: USDA, Forest Service, Intermountain Research Station, Intermountain Fire Sciences Laboratory. http://www.fs.fed.us/database/feis/plants/Shrub/ARCVIS. Accessed July 18, 1996.

Randall, W.R., R.F. Keniston, D.N. Bever, and E.C. Jensen. 1994. *Manual of Oregon Trees and Shrubs.* Corvallis, OR: Oregon State University Bookstores. 305p.

Artemisia tridentata
Big sagebrush

References

Deitschman, G.H. 1974. *Artemisia* L. Sagebrush. pp. 235-37 *In*: Schopmeyer, C.S. (tech. coord.) 1974. *Seeds of the Woody Plants in the United States*. Agric. Handbook 450. Washington, DC: USDA Forest Service. 883p.

Elias, T.S. 1980. *The Complete Trees of North America Field Guide and Natural History*. New York: Van Nostrand Reinhold Company. 948p.

Kruckeberg, A.R. 1982. *Gardening with Native Plants of the Pacific Northwest*. Seattle, WA: University of Washington Press. 252p.

Description

Big sagebrush is a shrub or small tree that grows up to 6 m. The trunk is short with branches close to the ground. The bark shreds into long, flat, thin strips and is grayish-brown in color. The leaves are simple, alternate, evergreen to late deciduous, wedge shaped, usually with three blunt teeth, 1-4 cm long, and covered with gray to white hairs on both sides. They give off a strong odor when crushed. Flowers are produced in small heads containing three to twelve tiny flowers and are located on elongated branched clusters at the ends of the branchlets. The fruit is a dry, hard, flattened achene, broadest near the tip, and brown in color. The leaves, flowers, and fruit are eaten by the sage grouse. Squirrels, rabbits, and small rodents eat the leaves and seeds, and antelope, mule deer, elk, and mountain sheep browse the leaves and young twigs. The wood of big sagebrush is hard and dense, and is used for firewood (Elias 1990).

Habitat and Geographic Range

Big sagebrush can be found on plains, deserts, hills, and lower mountain slopes in a variety of soils from British Columbia to Baja California, and east to western Nebraska (Elias 1990). It ranges from 450 to 2450 m in elevation and flourishes in deep, rich, moist, alluvial loams (Sampson and Jespersen 1981).

Propagation

Seed: Big sagebrush flowers in September or October and the fruits mature a month or two later. Collect seed by shaking, hand stripping, or beating the bush. Clean by hammermilling, but fanning and screening helps to reduce impurities. Seeds with 8-12% moisture content stored in cloth or burlap sacks or metal containers in an unheated warehouse maintain good viability for up to two years. Optimum temperatures for germination range from 16 to 18°C. Sow on nursery beds in the fall or winter, and cover with 0.5 cm of soil and a light straw mulch. After one or two years, outplant in early spring (Deitschman 1974).

Seeds per kilogram: ~5,253,525-7,138,450 (Deitschman 1974)

Vegetative: Take cuttings from half-ripened terminal or lateral twigs. Young seedlings transplanted in the fall also establish themselves well (Kruckeberg 1982).

Sampson, A.W., and B.S. Jespersen. 1981. California Range Brushlands and Browse Plants. Berkeley, CA: University of California Division of Agricultural Sciences. California Agricultural Experiment Station. Extension Service. 162p.

Betula occidentalis
Water birch

Description

Water birch grows as a many-stemmed shrub or occasionally as a small tree that can reach heights up to 12 m. The bark is smooth, nearly black on young trees, turning reddish-brown with horizontal lenticels. The deciduous leaves are alternate, simple, 2-5 cm long, 1.5-2.5 cm wide, ovate to elliptic in shape, with doubly serrate margins. The female and male flowers appear as separate catkins on the same tree. The male catkins are 5-6.5 cm long and are borne terminally on the branchlets. The fruit is a winged samara enclosed in a cone, 2.5-3 cm long, consisting of three-lobed scales. The seed is very small and light. Water birch rarely reaches commercial size and is locally used for firewood and fence posts. Beavers use the stems for building dams and lodges, hummingbirds feed on the sap, many species of birds eat the seed, buds, and catkins, and it is an occasional browse for livestock and big game. Water birch has a dense root system that is excellent at stabilizing streambanks. This species also hybridizes freely with paper birch (*Betula papyrifera*) in some areas of its range (Elias 1980, Uchytil 1989).

Habitat and Geographic Range

Water birch ranges from southern Alaska to southern Manitoba, south to North Dakota, and west to southern California and New Mexico. It is not found in the Pacific Coast mountain ranges and is scarce in the central and northern Sierra Nevadas. It occurs at low to middle elevations, usually from 600 to 2900 m. It is a riparian plant and can be found growing on alluvial terraces, along streams, springs, or other waterways. It is very flood tolerant and occurs on a variety of soils (Elias 1980, Uchytil 1989).

Propagation

Seed: Plants mature early and usually start producing seed at ten to twelve years of age. The female catkins turn brown upon maturing in late summer or fall and seed is dispersed throughout fall and into winter. Collect by picking the strobiles while they are still somewhat green and spreading them out to dry. Remove the seed

References

Brinkman, K.A. 1974. *Betula* L. Birch. pp. 252-57 *In*: Schopmeyer, C.S. (tech. coord.) 1974. *Seeds of the Woody Plants in the United States.* Agric. Handbook 450. Washington, DC: USDA Forest Service. 883p.

Elias, T.S. 1980. *The Complete Trees of North America Field Guide and Natural History.* New York: Van Nostrand Reinhold Company. 948p.

Platts, W.S., C. Armour, G.D. Booth, M. Bryant, J.L. Bufford, P. Cuplin, S. Jensen, G.W. Lienkaemper, G.W. Minshall, S.B. Monsen, R.L. Nelson, J.R. Sedell, and J.S. Tuhy. 1987. Methods for evaluating riparian habitats with applications to management. USDA Forest Service. Gen. Tech. Rep. INT-221. 177p.

by flailing and shaking and separate by screening and fanning. Seed of most *Betula* spp. can be stored at 2-3°C with a 1-3% moisture content. Stratification is not necessary if seed has adequate light for germination. Sow in the fall at a depth of 2-5 mm and transplant when seedlings are one to two years old. Seedlings require shade for two to three months during their first summer. Seedling establishment after transplanting is excellent and growth rates are rapid (Brinkman 1974, Platts et al. 1987, Uchytil 1989).

Uchytil, R.J. 1989. *Betula occidentalis*. In: Fischer, William C. (comp.) The Fire Effects Information System [Monograph Online]. Missoula, MT: USDA Forest Service, Intermountain Fire Sciences Laboratory. http://www.fs.fed.us/database/feis/plants/Tree/BETOCC. Accessed February 12, 1997.

Ceanothus cuneatus
Buckbrush

References

Huxley, A., M. Griffiths, and M. Levy. 1992. *New Royal Horticultural Society Dictionary of Gardening.* Vol 1. London: Macmillan Press Ltd.

Reed, M.J. 1974. *Ceanothus* L. Ceanothus. pp. 284-90 *In*: Schopmeyer, C.S. (tech. coord.) 1974. *Seeds of the Woody Plants in the United States.* Agric. Handbook 450. Washington, DC: USDA Forest Service. 883p.

Sampson, A.W., and B.S. Jespersen. 1981. California Range Brushlands and Browse Plants. Berkeley, CA: University of California Division of Agricultural Sciences. California Agricultural Experiment Station. Extension Service. 162p.

Tirmenstein, D. 1989. *Ceanothus cuneatus. In*: Fischer, William C. (comp.) The Fire Effects Information System [Monograph Online]. Missoula, MT: USDA Forest Service, Intermountain Fire Sciences Laboratory. http://www.fs.fed.us/database/feis/plants/Shrub/CEACUN. Accessed March 26, 1997.

Description

Buckbrush, also called wedgeleaf ceanothus, hornbrush, and greasebrush, is an erect, medium-sized, evergreen shrub. The branches are rigid and thorny. The leaves are thick, opposite, spatulate to obovate shaped, usually rounded at the tip, and with entire margins. Flowers are white to lavender in short, dense, round umbellate clusters. The capsule is nearly round with noticeable horns on the back. Buckbrush is good for nutrient-deficient soils due to its ability to fix nitrogen. It is also excellent browse and cover for many wildlife species, especially deer (Sampson and Jespersen 1981, Tirmenstein 1989).

Habitat and Geographic Range

Buckbrush is found between 90 and 1200 m in elevation on dry exposed mountain slopes, ridges, and semi-arid valleys with well-drained soils (Sampson and Jespersen 1981). It grows from Mexico northward to California, western Nevada, Oregon, and into southern Washington (Tirmenstein 1989).

Propagation

Seed: Fruit ripens from April to June. Collect seed by tying a cloth bag over the clusters of green capsules; as the capsules split, the ripe seeds will be released into the bag. Separate the seed by screening and fanning. Germination is increased when seeds are soaked in hot water (70°C) for 12 hours prior to a cold stratification. Sow in December or January in flats at a depth of 1.3-2.5 cm. When several sets of true leaves have formed, transplant into 2- or 3-inch pots (Reed 1974).
Seeds per kilogram: ~79,365-123,460 (Reed 1974)
Vegetative: Buckbrush can be propagated by softwood or semi-ripe nodal cuttings. Take cuttings in the summer, treat with a 0.8% IBA solution, and stick in a damp, porous, sandy medium. Apply bottom heat and pot immediately when rooted (Huxley et al. 1992).

Ceanothus prostratus
Squaw carpet

Description

Squaw carpet is a low-growing evergreen shrub that grows up to 45 cm in height and forms a dense mat 0.5-2.5 m across. The leaves are opposite, round to broadly elliptical, thick and leathery, and the margins are finely serrate near the tip. The flowers are borne in umbels at the end of the current year's growth and are a deep to light blue, frequently becoming pinkish with age. The fruit is a roundish capsule with large wrinkled horns. Buds and new growth are browsed by deer (Sampson and Jespersen 1981). Once established, squaw carpet is a valuable species for slope stabilization and erosion control (Post 1989).

Habitat and Geographic Range

Squaw carpet occurs in ponderosa pine and red fir forests in the north Coast Range between 640 and 2400 m in elevation (Sampson and Jespersen 1981). It grows best on well-drained soils in partial shade (Post 1989).

Propagation:

Seed: Squaw carpet fruits ripen in July. Collect the seed by tying a cloth bag over the capsule clusters; when the capsules split, the ripe seeds will be released into the bag. Clean the seed by screening and fanning (Reed 1974). Place in hot water (82°C) and then allow to cool and soak for 24 hours. Following the hot water treatment, mix the seed with moist sand and place in a refrigerator. Once the seeds have swollen, sow in small containers and cover with 1.5 cm of soil. After the plants have formed a third pair of leaves, transplant to larger containers. The plants will be ready for outplanting after two years (Post 1989).

Seeds per kilogram: ~81,565-98,105 (Reed 1974)

Vegetative: Softwood or semi-ripe cuttings can be taken during the summer. Treat with 0.8% IBA and stick in a damp, porous, sandy mixture. Apply bottom heat and pot immediately when rooted (Huxley et al. 1992).

References

Huxley, A., M. Griffiths, and M. Levy. 1992. *New Royal Horticultural Society Dictionary of Gardening.* Vol 1. London: Macmillan Press Ltd.

Post, R.L. 1989. Squaw carpet (*Ceanothus prostratus*). Plants for the Lake Tahoe Basin. Soil Conservation Service, Nevada Cooperative Extension. Fact Sheet 89-69.

Reed, M.J. 1974. *Ceanothus* L. Ceanothus. pp. 284-90 *In*: Schopmeyer, C.S. (tech. coord.) 1974. *Seeds of the Woody Plants in the United States.* Agric. Handbook 450. Washington, DC: USDA Forest Service. 883p.

Sampson, A.W., and B.S. Jespersen. 1981. California Range Brushlands and Browse Plants. Berkeley, CA: University of California Division of Agricultural Sciences. California Agricultural Experiment Station. Extension Service. 162p.

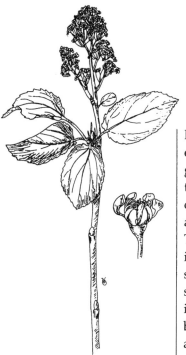

Ceanothus sanguineus
Redstem ceanothus

Description

Redstem ceanothus is a shade-intolerant, deciduous, erect shrub growing 1-3 m in height. The stems are greenish when young and darken to a purplish-red as they mature. The leaves are alternate, 2.5-9 cm long, thin, oval to elliptical in shape, dark green and glabrous above, paler below, three-veined, with serrate margins. The flowers are small, cream to white in color, and borne in a panicle. The fruits are three-celled capsules, with a single seed in each cell. Redstem ceanothus is a food source for elk, deer, hares, birds, rodents, and many insects. It also provides cover for many mammals and birds. Due to its deep root system and nitrogen-fixing ability, it is an excellent pioneer shrub for soil stabilization and improving soil fertility (Randall et al. 1994, Tirmenstein 1990).

Habitat and Geographic Range

Redstem ceanothus ranges from southern British Columbia to the Siskiyou Mountains in northern California, east to Idaho and western Montana. It grows best on moist, well-drained soils that are low in organics. It can be found at low to mid-elevations in the open or partial shade (Randall et al. 1994, Tirmenstein 1990).

Propagation

Seed: Sprouts of most species can produce at least some seed by three to six years of age. The seed of redstem ceanothus is approximately 2 mm in length, with a hard, impermeable seed coat and dormant embryo. Most species of ceanothus are prolific seed producers, although some annual variation has been noted (Tirmenstein 1990). Redstem ceanothus produces fruit during June and July; ripening is usually in August and appears to be associated with moisture stress. Collect seed just as the first capsules break open. Viable seed will turn dark brown as it matures while insect-infested or unfilled seed will be lighter in color. Collect capsules by hand and gently crack and break to prevent damage to the seed. Most of the cracked capsules will open if stored in dry covered containers for several days at 21-27°C. The seed can then be extracted with a seed cleaner.

References

Gratkowski, H. 1973. Pregermination treatments for redstem ceanothus seeds. Portland, OR: USDA Forest Service Pacific Northwest Forest and Range Experiment Station. Res. Pap. PNW-156. 10p.

Randall, W.R., R.F. Keniston, D.N. Bever, and E.C. Jensen. 1994. *Manual of Oregon Trees and Shrubs*. Corvallis, OR: Oregon State University Bookstores. 305p.

Reed, M.J. 1974. *Ceanothus* L. Ceanothus. pp. 284-90 *In*: Schopmeyer, C.S. (tech. coord.) 1974. *Seeds of the Woody Plants in the United States*. Agric. Handbook 450. Washington, DC: USDA Forest Service. 883p.

Store in dry paper containers at 3°C. Seed does not appear to lose much viability over long periods of time. Soak in 80-90°C water or dip in boiling water for ten seconds to five minutes; follow with a cold stratification for 90 days. Soaking seed in sulfuric acid for 30 minutes or dipping them in gibberellin will also help improve germination. Sow directly in the fall after heat treatment or pretreat and sow in spring (Gratkowski 1973, Tirmenstein 1990).

Seeds per kilogram: ~282,185-291,005 (Reed 1974)

Vegetative: Redstem ceanothus often sprouts after the crown is damaged or destroyed (Tirmenstein 1990).

Tirmenstein, D.A. 1990. *Ceanothus sanguineus. In*: Fischer, William C. (comp.) The Fire Effects Information System [Monograph Online]. Missoula, MT: USDA Forest Service, Intermountain Fire Sciences Laboratory. http://www.fs.fed.us/database/feis/plants/Shrub/CEASAN. Accessed June 25, 1996.

Ceanothus velutinus
Snowbrush ceanothus

References

Borland, J. 1988.
Ceanothus velutinus.
American Nurseryman.
168(9):154.

Huxley, A., M. Griffiths,
and M. Levy. 1992. *New*
Royal Horticultural Society
Dictionary of Gardening.
Vol 1. London: Macmillan
Press Ltd.

Reed, M.J. 1974.
Ceanothus L. Ceanothus.
pp. 284-90 *In*:
Schopmeyer, C.S. (tech.
coord.) 1974. *Seeds of the*
Woody Plants in the United
States. Agric. Handbook
450. Washington, DC:
USDA Forest Service.
883p.

Description

Snowbrush is a spreading evergreen shrub that can reach up to 6 m in height, but is more commonly 60-150 cm tall and wide. Leaves are alternate, broadly elliptical to elliptical-ovate, three-veined from the base, with finely toothed margins and a strong balsamlike odor. Young leaves have a shiny, varnished appearance while older leaves are dark green and feel somewhat velvety. The flowers are white, fragrant, and found in compound clusters about 5-10 cm in length. The fruit is a roundish to triangular capsule that is distinctly three-lobed at the top. The roots of snowbrush ceanothus have a nitrogen-fixing ability which is beneficial when this shrub is planted in nutrient-deficient soils. Snowbrush provides food and cover for a variety of wildlife species (Sampson and Jespersen 1981, Borland 1988).

Habitat and Geographic Range

Snowbrush grows in open wooded mountain slopes. It can be found from 1100 to 3000 m in elevation and ranges from British Columbia south to South Dakota, Colorado, Utah, Nevada, and California. It grows in many soil types but does best in deep, medium- to coarse-textured, well-drained soils. (Reed 1974, Sampson and Jespersen 1981, Borland 1988).

Propagation

Seed: Snowbrush ceanothus flowers bloom from May to June and the fruits ripen from July to September. The fruits explode when ripe, making it difficult to collect the seed. Collect by tying a cloth bag over the clusters of capsules; when they split, the seeds will be released into the bag. Separate the seed by screening and fanning. Germination is increased when seeds are soaked in hot water for several hours prior to a moist, cool (1-5°C) stratification for 60-85 days. Sow pretreated seed 0.5-1.5 cm deep in December or January in flats and transplant into 2- or 3-inch pots after several true leaves have formed. Emergent seedlings are susceptible to damping-off (Reed 1974, Borland 1988).

Seeds per kilogram: ~135,360-335,100 (Reed 1974)

Vegetative: Snowbrush can be propagated by softwood or semi-ripe nodal cuttings. Take cuttings in the summer, treat with 0.8% IBA, and stick in a damp, porous, sandy medium. Apply bottom heat and pot immediately when rooted (Huxley et al. 1992). Rooted cuttings can be difficult to overwinter and should be potted only in highly aerated soil (Borland 1988).

Sampson, A.W., and B.S. Jespersen. 1981. California Range Brushlands and Browse Plants. Berkeley, CA: University of California Division of Agricultural Sciences. California Agricultural Experiment Station. Extension Service. 162p.

Cercocarpus ledifolius
Curlleaf mountain-mahogany

Description

Curlleaf mountain-mahogany is an evergreen shrub or small tree which grows up to 12 m tall and 90 cm in diameter. It has a round, compact crown and a short trunk. The bark is reddish-brown and becomes deeply furrowed and scaly with age. The leaves are simple, alternate, lance shaped, 12-25 mm long, and leathery. The margins are entire and curled under. Leaves persist until the second season. The flowers are produced singly with a narrow calyx tube, no petals, twenty to thirty stamens, and a single pistil, and can be found in the junction of the leaves. The fruits are hard, rounded, narrow achenes, and are tipped with a persistent featherlike style. Deer use this species for browse and cover. The wood is an excellent source of fuel because it burns long and gives off intense heat (Elias 1980).

Habitat and Geographic Range

Curlleaf mountain-mahogany grows at elevations of 1500-2700 m on dry gravelly slopes. It ranges from northern Wyoming to southeastern Washington, south to southeastern California, and into western Colorado. It is found is association with lodgepole pine, aspen, firs, and spruces (Elias 1980).

Propagation

Seed: Seed production of this species can be sporadic. Seeds reach maturity during August and September. Pass the fruits through a hammermill to remove the styles. Seeds with a moisture content of 7-10% can be stored for up to five years in ventilated containers. Cold stratify before sowing or sow into seed flats and place in cold stratification, then transplant into growing containers. Use a growing medium that allows for adequate drainage to prevent root rot. Fertilization with nitrogen, phosphorus, and potassium at each watering is helpful for seedling growth (Link 1993). Seedlings are sensitive to drought and frost.

Seeds per kilogram: ~106,260-124,780 (Deitschman et al. 1974)

References

Deitschman, G.H., K.R. Jorgensen, and A.P. Plummer. 1974. *Cercocarpus* H.B.K. Cercocarpus (Mountain-mahogany). pp. 309-12 *In*: Schopmeyer, C.S. (tech. coord.) 1974. *Seeds of the Woody Plants in the United States.* Agric. Handbook 450. Washington, DC: USDA Forest Service. 883p.

Elias, T.S. 1980. *The Complete Trees of North America Field Guide and Natural History.* New York: Van Nostrand Reinhold Company. 948p.

Link, E. 1993. (ed.) Native Plant Propagation Techniques for National Parks: Interim Guide. East Lansing, MI: Rose Lake Plant Materials Center. 240p.

Cercocarpus montanus
(C. betuloides)
True mountain-mahogany

Description

True mountain-mahogany is a perennial shrub or small tree reaching up to 6 m in height with upright to spreading branches. The leaves are simple, alternate or somewhat fascicled, ovate to obovate, 2-5 cm long, 1.5-3.5 cm wide, with coarsely serrated margins. The stems are stout and roughened by leaf scars. The flowers are axillary and found solitary or in clusters of two to three. The fruit is an appressed-silky achene 8-10 mm long. Native Americans used the wood to make tools and the bark was used as a dye by the Hopi. True mountain-mahogany provides good erosion control in arid environments (Deitschman et al. 1974, Stubbendieck et al. 1992).

Habitat and Geographic Range

True mountain-mahogany ranges from Oregon east to Wyoming, into parts of South Dakota and Kansas, and south to southern California and central Mexico. Its elevational range is 1220-3050 m. It can be found growing on mountain sides, rocky bluffs, and rimrock, and in canyons and open woodland. It is most common on dry soils (Deitschman et al. 1974, USDA 1988, Stubbendieck 1992).

Propagation

Seed: Fruit ripens in late summer to early fall and is collected by shaking onto canvas bags or into hoppers. Clean by fanning and screening, or for easier planting run through a hammermill. Seed with 7-10% moisture content can be stored in wooden or metal containers or in cloth sacks in a dry, ventilated place for five years or more. Seed must be stratified to overcome dormancy. Sow stratified seed in the spring and unstratified seed during the fall. Keep seedbeds moist until germination begins. Outplant after one to two years (Deitschman et al. 1974).

Seeds per kilogram: ~123,235-143,740 (Deitschman et al. 1974)

References

Deitschman, G.H., K.R. Jorgensen, and A.P. Plummer. 1974. *Cercocarpus* H. B. K. Cercocarpus (Mountain-mahogany). pp. 309-12 *In*: Schopmeyer, C.S. (tech. coord.) 1974. *Seeds of the Woody Plants in the United States*. Agric. Handbook 450. Washington, DC: USDA Forest Service. 883p.

Stubbendieck, J., S.L. Hatch, and C.H. Butterfield. 1992. *North American Range Plants*. Fourth Edition. Lincoln, NE: University of Nebraska Press. 493p.

USDA Forest Service. 1988. *Range Plant Handbook*. New York: Dover Publications, Inc. 816p.

Cornus stolonifera (C. sericea)
Red-osier dogwood

Description

Red-osier dogwood is a deciduous, many-stemmed shrub that grows 2-5 m tall. The oppositely branched stems are smooth and purple to bright red in color, especially during dormancy. The leaves are 7-12 cm long, elliptic to ovate shaped with distinct, curving veins. The small white flowers are found in a flat-topped cyme and produce a white or bluish drupe. Red-osier dogwood is a favorable winter browse and forage for a variety of wildlife. The fruit is eaten by many different birds, and the twigs and foliage are browsed by whitetail deer, mule deer, snowshoe hare, moose, and elk (Elias 1980, Haeussler et al. 1990, King County 1994).

Habitat and Geographic Range

Red-osier dogwood has a wide range and an ability to tolerate extremely cold temperatures. It can be found at elevations of 450-2700 m from central Alaska to Newfoundland, south to California, and east to Nebraska and New York. It is a characteristic species of swamps, low meadows, and riparian zones. It is also found in forest openings, open forest understories, and along forest margins. It prefers wet sites and is usually found along streams in association with alders and willows (Elias 1980, Crane 1989).

Propagation

Seed: Plants first bear fruit at three to four years of age although older plants are more prolific (Crane 1989). The best time to collect the fruit is in August and September. This can be done easily by stripping or shaking the branches. Plant the drupes immediately into flats, or clean and process for later sowing. Remove the seed by macerating the fruit in water and floating away the pulp or running it through a hammermill. Dried seeds can be stored in sealed containers at 3-5°C for two to four years. Seed does best when sown in the fall and should be cold stratified at 3-5°C for 60 to 90 days prior to spring planting. Sometimes hard seed coats are present and scarification is necessary. Not all seeds will germinate right away. Plant into gallon containers or nursery beds, covered with 0.5-1.5 cm of soil and mulched with 1.5-2.5 cm of sawdust (Brinkman 1974).

References

Brinkman, K.A. 1974. *Cornus* L. Dogwood. pp. 336-42 *In*: Schopmeyer, C.S. (tech. coord.) 1974. *Seeds of the Woody Plants in the United States*. Agric. Handbook 450. Washington, DC: USDA Forest Service. 883p.

Buis, S. 1996. Owner, Sound Native Plants, Olympia, WA. Personal communication.

Crane, M.F. 1989. *Cornus sericea*. *In*: Fischer, William C. (comp.) The Fire Effects Information System [Monograph Online]. Missoula, MT: USDA Forest Service, Intermountain Fire Sciences Laboratory. http://www.fs.fed.us/database/feis/plants/Tree/CORSER. Accessed June 27, 1996.

Seeds per kilogram: ~30,420-58,865 (Brinkman 1974)

Vegetative: On good sites, red-osier dogwood forms dense thickets by layering when the lower stems touch the ground and root at the nodes (Crane 1989). Take cuttings of 5-8 cm from the branch tips in late summer, or whips of 0.5-1 m from from one-year-old wood during November through March when the tree is dormant (King County 1994). A mild-strength rooting hormone can be used though it is probably not necessary. Stick the cuttings in a perlite:vermiculite (1:1) mixture and set on a mist bench at 21°C. Once rooting has occured, transplant into a regular potting soil and move to a greenhouse (Buis 1996).

Elias, T.S. 1980. *The Complete Trees of North America Field Guide and Natural History.* New York: Van Nostrand Reinhold Company. 948p.

Haeussler, S., D. Coates, and J. Mather. 1990. Autecology of common plants in British Columbia: A literature review. British Columbia Ministry of Forests. FRDA Report-158. 272p.

King County Department of Public Works, Surface Water Management Division. 1994. Northwest Native Plants, Identification and Propagation for Revegetation and Restoration Projects. King County, WA. 68p.

Crataegus douglasii
Douglas hawthorn

References

Brinkman, K.A. 1974. *Crataegus* L. Hawthorn. pp. 356-60 *In*: Schopmeyer, C.S. (tech. coord.) 1974. *Seeds of the Woody Plants in the United States*. Agric. Handbook 450. Washington, DC: USDA Forest Service. 883p.

Buis, S. 1997. Owner, Sound Native Plants, Olympia, WA. Personal communication.

Elias, T.S. 1980. *The Complete Trees of North America Field Guide and Natural History*. New York: Van Nostrand Reinhold Company. 948p.

Habeck, R.J. 1991. *Crataegus douglasii. In*: Fischer, William C. (comp.) The Fire Effects Information System [Monograph Online]. Missoula, MT: USDA Forest Service, Intermountain Fire Sciences Laboratory. http://www.fs.fed.us/database/feis/plants/Tree/CRADOU. Accessed June 25, 1996.

Description

Douglas hawthorn, also known as black hawthorn, is a shrub or small tree usually about 4 m in height. The stems typically grow clustered near the base of the tree or just above the soil. Lower shade-killed limbs remain on the stem, forming large, dense thickets. The branches bear thorns up to 2.5 cm long. The leaves are oval, 3-6 cm long, and are irregularly lobed or serrated. The flowers are white, 1.5 cm in diameter, and are produced in clusters of five to twelve during the spring. The fruit is a dark purple to black berry about 1 cm in diameter. Douglas hawthorn provides food and both thermal and hiding cover for many wildlife species. Birds also use the thickets for nesting and brooding cover, and it is the preferred nesting habitat for magpies. Douglas hawthorn is an excellent soil and streambank stabilizer but does not typically occupy disturbed sites (Elias 1980, Habeck 1991, Strickler 1993).

Habitat and Geographic Range

Douglas hawthorn ranges from southeastern Alaska south to northern California and eastward to Colorado and western Montana. It can be found at lower elevations from 670 to 1645 m. It grows in wetland, riparian, or upland areas but can also be found on steep, uncultivated slopes and prefers deep, moist, fine-textured soils (Elias 1980, Habeck 1991).

Propagation

Seed: Seed ripens from late July through August and can be collected by gathering off the tree or off the ground. Frequent cutting tests are necessary during fruit collection since the number of sound seed can vary among trees. Extract seeds immediately either by macerating and floating off the pulp or by spreading them out in thin layers to avoid excess heating. Seeds should be thoroughly air dried before storing and can be kept for two to three years at 5°C. Douglas hawthorn requires a cold stratification of 5°C for 84-112 days. Brinkman (1974) reports that acid scarification on dry seeds for 0.5 to 3.0 hours prior to cold stratification will increase germination. On the other hand, Buis (1997) reports good germination with cold stratification only. Seed will generally yield 50 to 80% germination. Sow seed early in the fall or keep in moist, cool storage after acid scarification in the winter, and sow the following spring. Cover sown seeds with 0.5 cm of firm soil; they should not remain in the seedbed for longer than one year due to development of a long taproot. Successful establishment of seedlings is difficult and growth rates are slow (Brinkman 1974, Habeck 1991).

Seeds per kilogram: ~47,395-52,250 (Brinkman 1974)

Vegetative: Removal of aboveground stems will result in resprouts and suckers from the root system (Habeck 1991). Douglas hawthorn can be propagated by suckers and layering (Keator 1994).

Keator, G. 1994. *Complete Garden Guide to the Native Shrubs of California*. San Francisco, CA: Chronicle Books. 314p.

Strickler, D. 1993. *Wayside Wildflowers of the Pacific Northwest*. Columbia Falls, MT: The Flower Press. 272p.

Gaultheria shallon
Salal

References

Buis, S. 1996. Owner, Sound Native Plants, Olympia, WA. Personal communication.

Dimock, E.J. II., W.F. Johnston, and W.I. Stein. 1974. *Gaultheria* L. Wintergreen. pp. 422-26 *In*: Schopmeyer, C.S. (tech. coord.) 1974. *Seeds of the Woody Plants in the United States*. Agric. Handbook 450. Washington, DC: USDA Forest Service. 883p.

Fraser, L., R. Turkington, and C.P. Chanway. 1993. The biology of Canadian weeds. 102. *Gaultheria shallon* Pursh. *Canadian Journal of Plant Science* 73:1233-47.

Tirmenstein, D. 1990. *Gaultheria shallon*. *In*: Fischer, William C. (comp.) The Fire Effects Information System [Monograph Online]. Missoula, MT: USDA Forest Service, Intermountain Fire Sciences Laboratory. http://www.fs.fed.us/database/feis/plants/Shrub/GAUSHA. Accessed March 26, 1997.

Description

Salal is an erect to spreading, clonal evergreen and grows 0.5-2.5 m tall with a broad, shallow root system. Twigs are reddish-brown with shredding bark. The leaves are alternate, ovate, 3-10 cm long, 3-5 cm wide, shiny, and leathery. The small, urn-shaped flowers are white to pink and hang in showy clusters on the terminal. The fruit is a purplish-black, 6-10 mm, depressed capsule covered with tiny hairs. Each capsule contains approximately 125 brown 1-mm seeds. The leaves, twigs, and fruit are all food sources for a variety of animals (Tirmenstein 1990, Fraser et al. 1993).

Habitat and Geographic Range

Salal is predominantly found at lower elevations due to its sensitivity to frost. It is found from southeastern Alaska and central British Columbia southward to southern California along the Pacific Coast and inland to the western slope of the Cascades and Coast Ranges. It prefers humid climates and mild temperatures and grows on a wide range of soil types. It is most commonly found in association with coastal western hemlock and Douglas-fir. Salal is tolerant of salt spray and grows well in partial shade (Tirmenstein 1990, Fraser et al. 1993).

Propagation

Seed: Collect the fruit from August to October when the berries are dark purple in color. Clean by maceration and repeated washings. Because the seed is so small, it can be screened through nylon pantyhose. It will remain viable for several years in cold, dry storage. Stratification is not necessary, but at least eight hours of light per day are required for optimal germination. Sow in the fall by scattering with a salt shaker on a peat and perlite (1:1) medium. Salal seedlings are very slow growing and may take two to three years to reach 8-13 cm in height. Bottom heat can enhance the growth rate (Buis 1996). Once plants are large enough to handle, transplant to larger pots and outplant the following spring (Dimock et al. 1974).

Seeds per kilogram: ~5,670,190-8,333,335 (Dimock et al. 1974)

Vegetative: Salal reproduces readily through layering,

rhizomes, root suckering, and sprouting from the stem base (Tirmenstein 1990). Take stem cuttings approximately 15 cm in length from current year's growth. Make two 2.5-cm-long scars at the base, cutting just into the cambium. Then dip in a rooting hormone, stick in fine perlite, and maintain at a temperature of 20°C until rooting occurs (Van Meter 1975).

Van Meter, M. 1975. Propagation of *Gaultheria shallon* (Salal). Comb. *Proceedings of the International Plant Propagation Society* 25:77-78.

Holodiscus discolor
Oceanspray

Description

Oceanspray is a deciduous shrub that can grow up to 6 m tall but is generally 1-3.5 m tall. It has an erect to spreading form with many slender, arching branches. Leaves are alternate, oval to triangular shaped, prominently veined, and have many small lobes with fine teeth. The small cream-colored flowers grow numerously on terminal panicles that can reach up to 30 cm in length. The flower clusters turn brown and persist on the plant through the winter providing seed for birds and small mammals. The fruit is a small one-seeded achene. Oceanspray has good soil-binding characteristics. This shrub is also referred to as "ironwood" due to the strength and hardness of the wood (Helliwell 1987, McMurray 1987, Strickler 1993, King County 1994, Pojar and MacKinnon 1994).

Habitat and Geographic Range

Oceanspray ranges from British Columbia to western Montana, west to Oregon and south into southern California. It ranges in elevation from sea level to over 2100 m. It grows in upland, dryish open woods, canyon bottoms, sites along creekbanks and riverbanks, in forest openings and roadsides, and disturbed sites. It prefers shallow, rocky, well-drained soils but occupies a range of parent materials as well as soil textures. Oceanspray is considered a climax species in open conifer stands as well as an early seral species due to its ability to survive burning (McMurray 1987, Strickler 1993, King County 1994).

Propagation

Seed: Oceanspray typically blooms during late July. The fruit ripens in late August, and seed should be collected from September through November. Production of viable seed can be less than 10%. Oceanspray produces numerous small achenes which exhibit a pronounced dormancy. For optimum germination, seed must either be sown in the fall or go through a cold, moist stratification at 4-5°C for 15 to 18 weeks. Following stratification, seed will germinate if kept at temperatures of 20-24°C (Stickney 1974, McMurray 1987, Flessner et al. 1992, King Co. 1994).

References

Anderson, J. 1996. Owner. Sevenoaks Native Nursery, Corvallis, OR. Personal communication.

Flessner, T.R, D.C. Darris, and S.M. Lambert. 1992. Seed source evaluation of four native riparian shrubs for streambank rehabilitation in the Pacific Northwest. pp. 155-62 *In*: Proceedings, Symposium on Ecology and Management of Riparian Shrub Communities. USDA Forest Service Gen. Tech. Rep. INT-289.

Helliwell, R. 1987. Forest Plants of the Warm Springs Indian Reservation. Warm Springs, OR: Confederated Tribes of the Warm Springs. 177p.

King County Department of Public Works, Surface Water Management Division. 1994. Northwest Native Plants, Identification and Propagation for Revegetation and Restoration Projects. King County, WA. 68p.

Seeds per kilogram: ~11,772,485 (Stickney 1974)
Vegetative: Softwood cuttings do not work well. Hardwood cuttings can be collected in late January to early February when the plant drops its leaves (Anderson 1996) . Oceanspray can reproduce by perennating buds located on the root crown (McMurray 1987).

McMurray, N.E. 1987. *Holodiscus discolor. In*: Fischer, William C. (comp.) The Fire Effects Information System [Monograph Online]. Missoula, MT: USDA Forest Service, Intermountain Fire Sciences Laboratory. http://www.fs.fed.us/database/feis/plants/Shrub/HOLDIS. Accessed June 27, 1996.

Pojar, J., and A. MacKinnon. 1994. *Plants of the Pacific Northwest Coast: Washington, Oregon, British Columbia, and Alaska.* Vancouver, BC, Canada: British Columbia Ministry of Forests and Lone Pine Publishing. 527p.

Strickler, D. 1993. *Wayside Wildflowers of the Pacific Northwest.* Columbia Falls, MT: Flower Press. 272p.

Stickney, P.F. 1974. *Holodiscus discolor* (Pursh) Maxim. Creambush rockspiraea. pp. 448-49 *In*: Schopmeyer, C.S. (tech. coord.) 1974. *Seeds of the Woody Plants in the United States.* Agric. Handbook 450. Washington, DC: USDA Forest Service. 883p.

Lonicera involucrata
Bearberry honeysuckle

Description
Bearberry honeysuckle, also called black twinberry, is a shrub that grows 1-3 m high. Stems are four-sided when young, and leaves are opposite, elliptical, with arcunate veins, and entire margins. The yellow flowers are tubular and occur in pairs above prominent purple or green bracts. The fruit is a berry, about 6 mm in diameter, and reddish-black to black in color. The berries are often eaten by birds (Haeussler et al. 1990). Bearberry honeysuckle is a good soil binder.

Habitat and Geographic Range
Bearberry honeysuckle ranges from southern Alaska south to Mexico from lowlands to high elevations (Strickler 1993, Taylor and Douglas 1995). It grows in moist, open sites, such as wetlands and riparian areas and is also found in the forest understory (King County 1994, Buis 1996).

Propagation
Seed: The fruit ripens in July and August and can be collected by hand picking or stripping from the branches. Clean the seed by macerating and allowing the pulp and empty seed to float away. Air-dried seeds can be stored in sealed containers at 1-3°C for up to 15 years. Cold stratification for a period of 45 to 60 days increases germination rates. Sow in the fall and cover with a thin layer of soil and 5-7 cm of straw mulch (Brinkman 1974). Honeysuckle seedlings are susceptible to aphids (Buis 1996).

References

Buis, S. 1996. Owner, Sound Native Plants, Olympia, WA. Personal communication.

Brinkman, K.A. 1974. *Lonicera* L. Honeysuckle. pp. 515-19 *In*: Schopmeyer, C.S. (tech. coord.) 1974. *Seeds of the Woody Plants in the United States.* Agric. Handbook 450. Washington, DC: USDA Forest Service. 883p.

Haeussler, S., D. Coates, and J. Mather. 1990. Autecology of common plants in British Columbia: A literature review. British Columbia Ministry of Forests. FDRA Report 158. 272p.

Jebb, T. 1995. Horticulturalist, USDI Bureau of Land Management, C.A. Sprague Seed Orchard, Merlin, OR. Personal communication.

Seeds per kilogram: ~500,440-1,051,590 (Brinkman 1974)
Vegetative: Bearberry honeysuckle propagates easily from hardwood cuttings. Take cuttings during the dormant season and keep thoroughly moist over summer (King County 1994). Then dip in a rooting hormone and stick into containers with a growing medium of peat and perlite (1:1) (Jebb 1995) or potting soil (Buis 1996). Cuttings need plenty of water during their first summer.

King County Department of Public Works, Surface Water Management Division. 1994. Northwest Native Plants, Identification and Propagation for Revegetation and Restoration Projects. King County, WA. 68p.

Strickler, D. 1993. *Wayside Wildflowers of the Pacific Northwest.* Columbia Falls, MT: Flower Press. 272p.

Taylor, R.J,. and G.W. Douglas. 1995. *Mountain Plants of the Pacific Northwest: A Field Guide to Washington, Western British Columbia, and Southeastern Alaska.* Missoula, MT: Mountain Press Publishing Company. 437p.

Mahonia aquifolium
(Berberis aquifolium)
Shining Oregon-grape

References

Huxley, A., M. Griffiths, and M. Levy. 1992. *New Royal Horticultural Society Dictionary of Gardening.* Vol 1. London: Macmillan Press Ltd. 334p.

Keator, G. 1994. *Complete Garden Guide to the Native Shrubs of California.* San Francisco, CA: Chronicle Books. 314p.

Peck, M.E. 1961. *A Manual of the Higher Plants of Oregon.* Portland, OR: Binfords & Mort Publishers. 936p.

Rudolf, P.O. 1974. *Berberis* L. Barberry, mahonia. pp. 247-51 *In*: Schopmeyer, C.S. (tech. coord.) 1974. *Seeds of the Woody Plants in the United States.* Agric. Handbook 450. Washington, DC: USDA Forest Service. 883p.

USDA Forest Service. 1988. *Range Plant Handbook.* New York: Dover Publications, Inc. 816p.

Description

Shining Oregon-grape is an evergreen shrub 5-30 dm high with nearly erect branches. The leaves are compound with five to seven lance-ovate to lanceolate shaped leaflets, 5-10 cm long, with spiny-dentate margins. The flowers are racemes clustered at the ends of the branches. The fruit is a deep blue glaucous berry 8-9 mm long. Shining Oregon-grape is an occasional winter browse of deer and elk (Peck 1961, USDA 1988).

Habitat and Geographic Range

Shining Oregon-grape grows in woods and thickets (Peck 1961) and is moderately drought tolerant (Keator 1994). It ranges from western Oregon to Washington and British Columbia from sea level to the summit of the Cascade Mountains (Peck 1961, USDA 1988).

Propagation:

Seed: The fruit turns a deep glaucous blue when ripe, from late July to early August. Collect by hand picking or flail onto cloths or receptacles. Extract and clean the seeds by macerating with water and screening out or floating off the pulp. Dry the seeds superficially and sow immediately or store in sealed containers at temperatures slightly above freezing. Stored seeds should be cold stratified for 90 days before planting in the spring. Cover cleaned seeds sown in the fall with 0.3-1.3 cm of soil plus 0.5 cm of sand and mulch until germination begins. Outplant seedlings after two years (Rudolf 1974).

Seeds per kilogram: ~83,770-94,800 (Rudolf 1974)

Vegetative: Heeled, nodal, and basal cuttings can be taken into late autumn. Remove spines and leaves. Treat the proximal end with a rooting hormone and stick into a 2:1 vermiculite:sand medium in a cold frame (Huxley et al. 1992).

Disease: *Berberis* can become infected with a bacterium that causes small, purple-black spots on the leaves. This may progress to cause yellowing and premature leaf fall. It can be controlled by a protectant fungicide spray. *Berberis* can also be affected by armillaria root rot and powdery mildew (Huxley et al. 1992).

Mahonia nervosa
(Berberis nervosa)
Oregon-grape

Description
Oregon-grape, also called dwarf Oregon-grape or longleaf hollygrape, is a low-growing, rhizomatous, evergreen shrub that reaches 80 cm in height. The stems are simple, ascending to erect and generally grow in loose groups of several stems. The leaves are pinnately compound and borne in terminal tufts with seven to twenty-one dark green, glossy, leathery leaflets. The leaflets are ovate to lance-ovate or acute in shape with coarsely serrate to spinose margins. The yellow flowers are found on erect racemes that grow up to 21 cm long. The fruit is a large, dark blue, globose berry, 8-10 mm in diameter containing a number of black seeds. Oregon grape is browsed by black-tailed deer, the foliage is eaten by various small mammals, and the fruit are eaten by many small birds and mammals (USDA 1988, Tirmenstein 1990).

Habitat and Geographic Range
Oregon-grape ranges from southern British Columbia south to central California and occurs west of the Cascade Ranges and the Sierra Nevada. It can be found at low to mid-elevations. It grows on a variety of soils including coarse, shallow, rocky soils, coarse alluvium, or glacial outwash. Soils range from well to poorly drained and habitats vary from submontane to montane forests. It is a shade-tolerant species and can grow in the understory of 300-600-year-old forests, but also grows in open meadows and recent clearcuts (Tirmenstein 1990).

Propagation
Seed: Oregon-grape flowers in early to late spring and fruit ripens during July and August. Collect fruit in August to September using heavy gloves or flail onto cloths or receptacles spread beneath the bushes. Clean the seed by macerating with water and allowing the pulp to float off. Sow immediately in 1.25 cm of soil plus 0.6 cm of sand and stratify over winter, or dry and store in sealed containers at temperatures just above freezing. Spring-sown seeds should be cold stratified for six weeks at 4°C. Outplant after two years (Rudolf 1974, Foster 1977).

References
Buis, S. 1996. Owner, Sound Native Plants, Olympia, WA. Personal communication.

Foster, C.O. 1977. *Plants-a-Plenty: How to Multiply Outdoor and Indoor Plants Through Cuttings, Crown and Root Divisions, Grafting, Layering, and Seeds.* Emmaus, PA: Rodale Press, Inc. 328p.

Huxley, A., M. Griffiths, and M. Levy. 1992. *New Royal Horticultural Society Dictionary of Gardening.* Vol 1-4. London: Macmillan Press Ltd. 3240p.

Rudolf, P.O. 1974. *Berberis* L. Barberry, mahonia. pp. 247-51 *In*: Schopmeyer, C.S. (tech. coord.) 1974. *Seeds of the Woody Plants in the United States.* Agric. Handbook 450. Washington, DC: USDA Forest Service. 883p.

Tirmenstein, D.A. 1990. *Mahonia nervosa. In*: Fischer, William C. (comp.) The Fire Effects Information System [Monograph Online]. Missoula, MT: USDA Forest Service, Intermountain Fire Sciences Laboratory. http://www.fs.fed.us/database/feis/plants/Shrub/BERNER. Accessed July 18, 1996.

USDA Forest Service. 1988. *Range Plant Handbook*. New York: Dover Publications, Inc. 816p.

Seeds per kilogram: ~50,705-66,140 (Rudolf 1974)

Vegetative: Oregon-grape is rhizomatous and gradually expands laterally in the absence of disturbance. Plants generally sprout from rhizomes after aboveground portions of the plant are destroyed. Vegetative regeneration appears to be the dominant mode of regeneration after fire or other disturbances (Tirmenstein 1990). Oregon-grape can be propagated by taking heeled, nodal, and basal cuttings. Remove spines and leaves and treat the base with a rooting hormone. Stick cuttings in a 2:1 vermiculite:sand mixture in a cold frame (Huxley et al. 1992). Buis (1996) does not recomend cuttings due to their bulkiness and difficulties getting them to stay in the flats, but has had moderate success with root cuttings on nursery stock or salvaged plants. However, root cuttings can damage the parent plant.

Mahonia repens
Creeping Oregon-grape

Description

Creeping Oregon-grape is a low-growing, evergreen shrub that grows from long branching rootstalks. The stems arise from a rhizome and are commonly 10-30 cm in height. Roots can extend up to 1.8 m. The leaves are pinnately compound with three to seven coarsely serrated leaflets, ovate to elliptical in shape. The flowers are yellow and grow in an erect raceme. The fruit is a globose to oblong dark blue berry borne in grapelike clusters. The berries are eaten by a variety of birds and mammals and the plants are a winter food source for deer and elk, though Creeping Oregon-grape may be poisonous to livestock. It is a good species for revegetating game rangelands, old mines, roadsides, and recreation areas. It is also used for landscaping because it provides a ground cover that is both heat and drought tolerant. Dyes were made from the roots and berries by the Navajo, and the Great Basin tribes made a tea from the roots to cure many ailments (Peck 1961, Helliwell 1987, Walkup 1991).

Habitat and Geographic Range

Creeping Oregon-grape ranges throughout the western United States from western Texas to Arizona and California north to British Columbia and Alberta. It occurs from near sea level to 3050 m. It can be found on wooded slopes, shaded hillsides, or occasionally on open hillsides. It grows in sandy loams to silts, sandstones, and sedimentary shales. It is shade tolerant but also grows well in full sunlight. Creeping Oregon-grape is considered a climax dominant species (Walkup 1991).

Propagation

Seed: Creeping Oregon-grape flowers from early May through early August and fruit ripens from late June to mid-September. Collect the seed by hand stripping into hoppers and clean by macerating with water and floating the pulp off the top. After the seed has dried, remove the remaining debris with a fanning mill. Seed can be stored for up to five years in sealed containers kept slightly above 0°C. Stratify moist at 1°C for 30 days,

References

Helliwell, R. 1987. Forest Plants of the Warm Springs Indian Reservation. Warm Springs, OR: Confederated Tribes of the Warm Springs. 177p.

Peck, M.E. 1961. *A Manual of the Higher Plants of Oregon.* Portland, OR: Binfords & Mort Publishers. 936p.

Plummer, A.P., D.R. Christensen, and S.B. Monsen. 1968. Restoring Big Game Range in Utah. Utah Div. Fish Game Publ. 68-3. 183p.

Rudolf, P.O. 1974. *Berberis* L. Barberry, mahonia. pp. 247-51 *In*: Schopmeyer, C.S. (tech. coord.) 1974. *Seeds of the Woody Plants in the United States.* Agric. Handbook 450. Washington, DC: USDA Forest Service. 883p.

Vories, K.C. 1980. Growing Colorado Plants from Seed: A State of the Art. Vol. 1: Shrubs. Ogden, UT: USDA Forest Service Intermountain Forest and Range Experiment Station. Gen. Tech. Rep. INT-103. 80p.

Walkup, C.J. 1991. *Mahonia repens. In*: Fischer, William C. (comp.) The Fire Effects Information System [Monograph Online]. Missoula, MT: USDA Forest Service, Intermountain Fire Sciences Laboratory. http://www.fs.fed.us/ database/feis/plants/ Shrub/MAHREP. Accessed February 7, 1997.

then warm stratify at 20°C for 60 days, followed by another moist chill at 1°C for 196 days. Alternatively, cold stratify at 2°C for 16 weeks in a solution of gibberellic acid. Plant in well-drained soil and cover with a thin layer of soil and sand. Creeping Oregon-grape can also be direct seeded in the fall at a rate of 100-200 seeds per square meter (Plummer et al. 1968, Vories 1980, Walkup 1991).

Seeds per kilogram: ~119,045-156,525 (Rudolf 1974)

Vegetative: Creeping Oregon-grape can be propagated by suckers, cuttings, and layering (Walkup 1991).

Oplopanex horridum
Devil's club

Description

Devil's club is an erect to spreading deciduous shrub growing 1-3 m tall. The stems are thick, sparsely branched, and armed with dense yellowish spines up to 1 cm long. Leaves are alternate, up to 35 cm across, cordate shaped, with sharply serrated margins, five to thirteen lobes, and many spines on the underside. The flowers are small, whitish, arranged in dense pyramid-shaped terminal clusters. The fruits are bright red, flattened berries containing two or three seeds. Plants growing along streams provide shade cover for salmonoid fishes. Bears eat the berries, leaves, and stems. Devil's club is an important medicinal plant throughout its range for aboriginal peoples and almost every part of the plant is of use (Peck 1961, Howard 1993, Pojar and MacKinnon 1994).

Habitat and Geographic Range

Devil's club ranges from Alaska to southwest Oregon, east to western Montana and northern Idaho. There are also local populations on several islands in northern Lake Superior. It can be found in moist woods, along streams, and drainage areas from sea level to 1525 m. It is shade tolerant and grows in well-drained to poorly drained soils with sandy, silty, or loamy textures. It is indicative of late seral, climax, or old-growth forests and is often found growing in western redcedar and western hemlock understories (Howard 1993, Strickler 1993).

Propagation

Seed: Flowers from late spring to midsummer; fruits ripen approximately four weeks after flowering and persist through winter (Howard 1993).

Vegetative: In late spring to early summer, cut long stems from the sprawling horizontal branches, place into paper bags and keep moist. Cut the stem into 13-cm sections containing at least one bud scale scar. The use of a rooting hormone has not improved rooting. Place cuttings horizontally in well-drained potting mix and perlite with half the diameter above the soil. The end distal to the bud scale scar typically rots. Cuttings from the vertical stems stuck in a standard potting mix with perlite will also grow roots (Roorbach 1997).

References

Howard, J.L. 1993. *Oplopanax horridus. In*: Fischer, William C. (comp.) The Fire Effects Information System [Monograph Online]. Missoula, MT: USDA Forest Service, Intermountain Fire Sciences Laboratory. http://www.fs.fed.us/database/feis/plants/Shrub/OPLHOR. Accessed March 4, 1997.

Peck, M.E. 1961. *A Manual of the Higher Plants of Oregon*. Portland, OR: Binfords & Mort Publishers. 936p.

Pojar, J., and A. MacKinnon. 1994. *Plants of the Pacific Northwest Coast: Washington, Oregon, British Columbia, and Alaska*. Vancouver, BC, Canada: British Columbia Ministry of Forests and Lone Pine Publishing. 527p.

Roorbach, A. 1997. Oregon State University. Personal communication.

Strickler, D. 1993. *Wayside Wildflowers of the Pacific Northwest*. Columbia Falls, MT: Flower Press. 272p.

Pachistima myrsinites
Oregon boxwood

Description
Oregon boxwood is a low-growing, many-branched, broadleaf evergreen shrub that grows 0.3-1 m tall. The leaves are opposite, 1-3 cm long, oblong and glabrous with serrated margins. The reddish to purple flowers are very small and are found in small clusters in the leaf axils. The fruit is a tiny, two-seeded, berrylike capsule. Oregon boxwood is browsed by deer, elk, moose, mountain sheep, and grouse (Helliwell 1987, Snyder 1991).

Habitat and Geographic Range
Oregon boxwood ranges from British Columbia south to California and Mexico and east through the Rocky Mountains. It can be found from sea level up to 3350 m in elevation. It is a climax species and tolerates both sun and shade. It grows on well-drained, shallow, gravelly clay and silt loam soils but prefers light to deep shade with a moist atmosphere (Snyder 1991).

Propagation:
Seed: Oregon boxwood fruits from June through September (Snyder 1991). The seed may remain viable for decades (Helliwell 1987).
Vegetative: The shrub is easily propagated through stem cuttings (Snyder 1991), which are taken in the early fall. The nodes are very close, and cuttings only need to be three to four nodes long. Make a slant cut at the bottom and strip the bottom leaves off. Dip the end in a liquid rooting hormone and stick into a perlite:vermiculite (1:1) growing medium. Set plants on a mist bench in a greenhouse at 20°C with low mist. Once rooted, transplant to a potting media of peat, perlite, and ground bark (Buis 1996). Oregon boxwood reproduces naturally by layering.

References
Buis, S. 1996. Owner, Sound Native Plants, Olympia, WA. Personal communication.

Helliwell, R. 1987. Forest Plants of the Warm Springs Indian Reservation. Warm Springs, OR: Confederated Tribes of the Warm Springs. 177p.

Snyder, S.A. 1991. *Pachistima myrsinites. In*: Fischer, William C. (comp.) The Fire Effects Information System [Monograph Online]. Missoula, MT: USDA Forest Service, Intermountain Fire Sciences Laboratory. http://www.fs.fed.us/database/feis/plants/Shrub/PACMYR. Accessed June 27, 1996.

Philadelphus lewisii
Mockorange

Description
Mockorange, also called Lewis' mockorange, is a deciduous shrub that grows 1.5-2.5 m in height. The leaves are opposite, ovate to elliptical, 2.5-7 cm long, with entire to few widely spaced serrations on the margin. The white, fragrant flowers have four petals and are borne singly on stems or in loose terminal clusters. The fruit is a light brown capsule. Mockorange is browsed by deer and elk and the seed is eaten by quail and squirrels (Helliwell 1987).

Habitat and Geographic Range
Mockorange is a riparian species and can be found in gullies, tallus slopes, rocky cliffs, or along streams at low elevations (Helliwell 1987). It is widely distributed from British Columbia and the Cascade Mountains of Oregon and Washington to northern California, and eastward to Montana. It can be found at elevations from sea level to 2100 m (Harris 1988).

Propagation
Seed: The fruit matures in late summer. Extract seeds by gently crushing the dried capsules and passing them through an aspirator. Stratify at 5°C for eight weeks followed by 22-26°C (Stickney 1974).
Seeds per kilogram: ~7,716,045-17,636,685 (Stickney 1974)
Vegetative: Take softwood cuttings in June and July, dip in 1000 ppm IBA solution, stick in a peat:perlite (1:1) medium, and mist. Collect hardwood cuttings in fall or spring to a length of 20 cm. Treat with a 2500-8000 ppm IBA solution and insert 15 cm deep into a sandy soil. Fall plantings should be mulched (Dirr and Heuser 1987).

References
Dirr, M.A., and C.W. Heuser, Jr. 1987. *The Reference Manual of Woody Plant Propagation: From Seed to Tissue Culture.* Athens, GA: Varsity Press. 239p.

Harris, H.T. 1988. *Philadelphus lewisii. In*: Fischer, William C. (comp.) The Fire Effects Information System [Monograph Online]. Missoula, MT: USDA Forest Service, Intermountain Fire Sciences Laboratory. http://www.fs.fed.us/database/feis/plants/Shrub/PHILEW. Accessed March 26, 1997.

Helliwell, R. 1987. Forest Plants of the Warm Springs Indian Reservation. Warm Springs, OR: Confederated Tribes of the Warm Springs. 177p.

Stickney, P.F. 1974. *Philadelphus lewisii* Pursh. Lewis mockorange. pp. 580-81 *In*: Schopmeyer, C.S. (tech. coord.) 1974. *Seeds of the Woody Plants in the United States.* Agric. Handbook 450. Washington, DC: USDA Forest Service. 883p.

Physocarpus capitatus
Pacific ninebark

References

Anderson, J. 1996. Owner. Sevenoaks Native Nursery, Corvallis, OR. Personal communication.

Buis, S. 1997. Owner. Sound Native Plants, Olympia, WA. Personal communication.

King County Department of Public Works, Surface Water Management Division. 1994. Northwest Native Plants: Identification and Propagation for revegetation and restoration projects. King County, WA. 68p.

Kruckeberg, A.R. 1982. *Gardening with Native Plants of the Pacific Northwest*. Seattle, WA: University of Washington Press. 252 p.

Pojar, J., and A. MacKinnon. 1994. *Plants of the Pacific Northwest Coast: Washington, Oregon, British Columbia, and Alaska*. Vancouver, BC, Canada: British Columbia Ministry of Forests and Lone Pine Publishing. 527p.

Strickler, D. 1993. *Wayside Wildflowers of the Pacific Northwest*. Columbia Falls, MT: Flower Press. 272p.

Van Dersal, W.R. 1938. Native Woody Plants of the United States: Their Erosion-Control and Wildlife Values. USDA Misc. Pub. 303. 362p.

Description

Pacific ninebark is a deciduous, erect to spreading shrub growing up to 4 m in height. The bark is reddish- to yellowish-brown and shreds in layers. The leaves are alternate, 3-6 cm long, with three to five serrated lobes and palmate veins. The white flowers are small with five petals and are borne in a rounded terminal cluster. The fruits are bunches of dried inflated reddish follicles about 1 cm long with shiny, yellowish seeds inside. Pacific ninebark is useful for preventing soil erosion (King County 1994, Pojar and MacKinnon 1994).

Habitat and Geographic Range

Pacific ninebark ranges from Alaska to California west of the Cascade Mountains with a small population in northern Idaho. It grows in moist, somewhat open places but can occasionally be found on drier, brushy sites. It is found from low to mid-elevations (Strickler 1993, Pojar and MacKinnon 1994).

Propagation

Seed: Pacific ninebark flowers from April to July. Seed can be collected from late August through September and sown in the fall (Van Dersal 1938, Kruckeberg 1982, King County 1994).

Vegetative: Hardwood cuttings root better than softwood cuttings. Take cuttings 15 cm long in mid-winter. Store in sawdust and stick in late winter or early spring in a sand medium (Anderson 1996). Dipping in a liquid rooting hormone can improve results (Buis 1997).

Physocarpus malvaceus
Mallow ninebark

Description
Mallow ninebark is a decidous shrub that grows 0.5-2 m tall. It has a broad stem structure with grayish bark that shreds on older stems. The leaves are alternate, palmately three- to five-lobed, and doubly serrated at the margins. Flowers are found in a tight, round cluster with five white petals and a brownish yellow center. The fruit is a swollen, two- to three-chambered capsule with two to four small seeds per chamber. Mallow ninebark is not an important browse species and is generally avoided. It provides shelter and cover for many wildlife species from small birds to large mammals when it forms dense thickets (Habeck 1992, Strickler 1993).

Habitat and Geographic Range
Mallow ninebark ranges east of the Cascade Mountains from south-central British Columbia to central Oregon, east to southwestern Alberta, Montana, Wyoming, Nevada, and Utah. It can be found at elevations of 1600-3000 m and grows on hillsides, canyons, and grasslands on mesic sites predominantly occupied by ponderosa pine and Douglas-fir. It is found mainly on soils with no exposed rock and with soil textures ranging from sandy loam to silty clay loam (Habeck 1992).

Propagation
Seed: Fruit ripens from August to early October when it turns from pale green to pale brown. Collect seeds by hand picking or shaking onto a dropcloth. Dry naturally or with artificial heat, then thresh and clean. Dried seeds can be stored at room temperature and will remain viable for at least five years. Sow in the fall and cover with mulch or stratify with a 30-day prechill and sow in the spring. Germination is generally low due to a large percentage of unsound seed (Gill and Pogge 1974, Habeck 1992, Link 1993).
Seeds per kilogram: ~1,666,670 (Gill and Pogge 1974)
Vegetative: Propagation of mallow ninebark is easier by cuttings than by seed (Gill and Pogge 1974). It can also be propagated using rhizomes. There can be numerous suppressed buds along the entire length of the rhizome axis that may permit sprouting at many points given the proper environmental conditions (Habeck 1992).

References
Gill, J.D., and F.L. Pogge. 1974. *Physocarpus* Maxim. Ninebark. pp. 584-86 *In*: Schopmeyer, C.S. (tech. coord.) 1974. *Seeds of the Woody Plants in the United States*. Agric. Handbook 450. Washington, DC: USDA Forest Service. 883p.

Habeck, R.J. 1992. *Physocarpus malvaceus*. In: Fischer, William C. (comp.) The Fire Effects Information System [Monograph Online]. Missoula, MT: USDA Forest Service, Intermountain Fire Sciences Laboratory. http://www.fs.fed.us/database/feis/plants/Shrub/PHYMAL. Accessed June 25, 1996.

Link, E. 1993. (ed.) Native Plant Propagation Techniques for National Parks: Interim Guide. East Lansing, MI: Rose Lake Plant Materials Center. 240p.

Strickler, D. 1993. *Wayside Wildflowers of the Pacific Northwest*. Columbia Falls, MT: Flower Press. 272p.

Potentilla fruticosa
Shrubby cinquefoil

References

Dirr, M.A., and C.W. Heuser, Jr. 1987. *The Reference Manual of Woody Plant Propagation: From Seed to Tissue Culture.* Athens, GA: Varsity Press. 239p.

Government of Saskatchewan. 1989. Guide to Forest Understory Vegetation in Saskatchewan. Canada Forestry. Technical Bulletin No. 9/1980 rev. January, 1989. 106p.

Link, E. 1993. (ed.) Native Plant Propagation Techniques for National Parks: Interim Guide. East Lansing, MI: Rose Lake Plant Materials Center. 240p.

Snyder, L.C. 1991. Native Plants for Northern Gardens. Anderson Horticultural Library, University of Minnesota. 277p.

Tirmenstein, D. 1987. *Potentilla fruticosa. In*: Fischer, William C. (comp.) The Fire Effects Information System [Monograph Online]. Missoula, MT: USDA Forest Service, Intermountain Fire Sciences Laboratory. http://www.fs.fed.us/ database/feis/plants/ Shrub/POTFRU. Accessed April 3, 1997.

Description
Shrubby cinquefoil is a brushy shrub that reaches 25-90 cm in height. Most of the branches grow from the rootstock. The leathery, persistent leaves are alternate, pinnately compound with five to seven oblong leaflets, and gray-green in color. The white to yellow flowers have five rounded petals, and are found singly or in terminal cymes. The fruit is a densely hairy achene which is light brown when mature (Tirmenstein 1987, Govt. of Saskatchewan 1989). This is not a valuable species for wildlife or livestock browsing, though small birds and mammals consume the seeds. Because of its rapid growth rate, good tolerance to cold, and ability to thrive on a range of soils, shrubby cinquefoil can be a good species for revegetation of disturbed sites (Tirmenstein 1987).

Habitat and Geographic Range
Shrubby cinquefoil grows well on a variety of sites. It ranges throughout North America from 1500 to 3500 m in elevation (Link 1993) and can be found on lower foothills, dry, gravelly areas, tallus, open mountain meadows, subalpine slopes and moist sites such as swamps, marshes, and streambanks.

Propagation
Seed: Seeds mature from July to September and can be cleaned with an air screen cleaner. Seed can be stored dry at 1-5°C and will remain viable for up to five years. Seeds require a 60-day stratification at 1°C and can be seeded directly at a depth of 4-6 mm (Link 1993). Germination percentages can be low (Tirmenstein 1987).
Seeds per kilogram: ~2,204,585 (Tirmenstein 1987)
Vegetative: Take softwood cuttings in July. Dip in 1000 ppm IBA and plant in a peat:perlite or sand medium and mist. Rooting occurs in about three weeks. Cuttings can then be transplanted or removed from the mist (Dirr and Heuser 1987, Snyder 1991). Shrubby cinquefoil also reproduces via layering (Tirmenstein 1987).

Purshia tridentata
Antelope bitterbrush

Description

Bitterbrush is a widely branched shrub 0.6-2.5 m tall with a silvery appearance. The leaves are simple, alternate, three-lobed at the apex, wedge shaped, 0.5-1.25 cm long, and hairy on both sides. The flowers are white or yellowish, found singly or in small groups at the ends of the branchlets. The fruit is a leathery, oblong achene, 0.6-1.25 cm long. It is an excellent food source for deer, sheep, cattle, squirrels, and chipmunks and can be used to reduce erosion on steep slopes and gullies (Sampson and Jespersen 1981, Post 1989).

Habitat and Geographic Range

Bitterbrush is distributed from British Columbia south along the east side of the Cascades and the Columbia Gorge in Washington and Oregon, south to California and east to western Montana, Wyoming, Colorado, and New Mexico (Bradley 1986). It is found on open, well-drained flats, slopes, and valleys with somewhat deep gravel or rocky soils at elevations between 60 and 3000 m. It grows in association with sagebrush, curlleaf mountain mahogany, ponderosa pine, western juniper, and various perennial grasses and forbs (Sampson and Jespersen 1981).

Propagation

Seed: As the achenes ripen in the late summer months, they change color from light or dark red to gray. Collect by placing canvasses or containers below the shrub and hitting the limbs or shaking the branches. Clean with a dewinger and separate from the husks and debris with a conventional fanning mill. Field-dried seed can be stored in burlap bags or airtight containers in a cool, dry place for up to five years (Deitschman et al. 1974). Stratify for two to seven weeks at 2-5°C to break dormancy (Meyer and Monsen 1989). Seed can be directly sown in the fall or spring, but stratified seed must be sown while wet. Do not plant deeper than 2 cm. If a short-term treatment is needed, seeds can be treated with a 3% solution of hydrogen peroxide for five hours to enhance germination (Young and Evans 1983, Young and Young 1986).

References

Bradley, A.F. 1986. *Purshia tridentata. In*: Fischer, William C. (comp.) The Fire Effects Information System [Monograph Online]. Missoula, MT: USDA Forest Service, Intermountain Fire Sciences Laboratory. http://www.fs.fed.us/database/feis/plants/Shrub/PURTRI. Accessed April 4, 1997.

Deitschman, G.H., K.R. Jorgensen, and A.P. Plummer. 1974. *Purshia* DC. Bitterbrush. pp. 686-88 *In*: Schopmeyer, C.S. (tech. coord.) 1974. *Seeds of the Woody Plants in the United States*. Agric. Handbook 450. Washington, DC: USDA Forest Service. 883p.

Kituku, V.M., W.A. Laycock, J. Powell, and A.A. Beetle. 1995. Propagating bitterbrush twigs for restoring shrublands. pp. 327-28 *In*: Roundy, B.A., E. McArthur, J.S. Haley, and D.K. Mann (comps.). Proceedings: Wildland Shrub and Arid Land Restoration Symposium. Las Vegas, NV; October 19-21, 1993. USDA Forest Service, Intermountain Research Station. INT-GTR-315.

Meyer, S.E,. and S.B. Monsen. 1989. Seed germination biology of antelope bitterbrush (*Purshia tridenta*). pp. 147-57 *In*: Proceedings, Symposium on Shrub Ecophysiology and Biotechnology, Logan,

UT. Wallace, A., E.D. McArthur, and M.R. Haferkamp (comp.). USDA Forest Service, Intermountain Res. Sta. Gen. Tech. Rep. INT-256.

Post, R.L. 1989. Antelope bitterbrush (*Purshia tridentata*). Plants for the Lake Tahoe Basin. Soil Conservation Service, Nevada Coop. Extension. Fact Sheet 89-45.

Sampson, A.W., and B.S. Jespersen. 1981. California Range Brushlands and Browse Plants. Berkeley, CA: University of California Division of Agricultural Sciences. California Agricultural Experiment Station. Extension Service. 162p.

Young, J.A., and R.A. Evans. 1983. Seed physiology of antelope bitterbrush and related species. pp. 70-80 *In*: Proceedings, Research and Management of Bitterbrush and Cliffrose in Western North America. Tiedmann, A.R., and K.L. Johnson (comp.). USDA Forest Service, Intermountain Forest and Range Experiment Station. Gen. Tech. Rep. INT-152.

Young, J.A., and C.G. Young. 1986. *Collecting, Processing and Germinating Seeds of Wildland Plants*. Portland, OR: Timber Press. 236p.

Seeds per kilogram: ~29,540-41,890 (Deitschman et al. 1974)

Vegetative: Bitterbrush can be propagated by stem cuttings collected in June. Take cuttings at least 10 cm long from current growth with a heel from older wood. Keep cool and moist during handling to avoid excessive water loss. Moisten heels in water then dip in a 0.3% IBA preparation. Stick cuttings 2-4 cm deep in a rooting medium of sand, pumice, and vermiculite (1:1:1). Keep the temperature 20-25°C. Irrigate daily until roots develop (45-60 days). After root development, transplant to containers (Kituku et al. 1995). In nature, bitterbrush plants with a decumbent growth form can reproduce via layering (Bradley 1986). This can be accomplished artificially by bending the branch down into a small hole so that the end protrudes above the soil surface. Anchor the branch to the bottom of the hole with a rock or wire and cover with soil. Secure the protruding end to a stake and keep the soil around the branch moist. After root development (which could take up to two years), new plants can be cut free from the parent plant and transplanted (Post 1989).

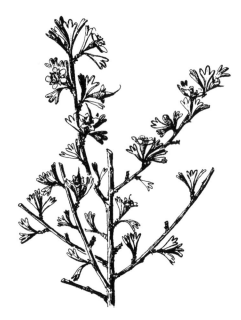

Rhamnus purshiana
Cascara buckthorn

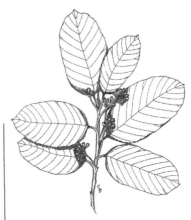

Description

Cascara buckthorn is a deciduous tall shrub or small tree. It can grow up to 10 m in height, but is generally smaller and bushier along its southern range. The bark is thin, smooth, developing thin scales with age, and silvery-gray in color. The leaves are simple, alternate, egg-shaped to oblong, 6-12 cm long, and distinctly pinnately veined in channels. The winter buds have no scales and are covered with dense hairs. The flowers are small, greenish-yellow, with five sepals, petals, and stamens, and a single pistil. They are borne eight to fifty in stalked umbels in the axils of the leaves. The fruit is a blue- to purplish-black drupe, 5-8 mm across, containing two or three seeds. Cascara buckthorn is a winter browse for mule deer and elk, and the drupes are eaten by many birds. It also provides thermal and hiding cover for many species of wildlife. The bark was used by natives as a laxative (Elias 1980, Habeck 1992, Pojar and MacKinnon 1994).

Habitat and Geographic Range

Cascara buckthorn ranges from southern British Columbia east to western Montana and southwest to central California. It grows in canyons, bottomlands, and lower mountain slopes up to 900 m in elevation. It is indicative of moist sites and is very tolerant of shade. It can be found growing with Douglas-fir, western redcedar, hemlock, and many maples (Hubbard 1974, Elias 1980, Habeck 1992).

Propagation

Seed: Cascara buckthorn flowers in May or June and the fruit ripens from July through September. Collect by picking fruit from the plant about two weeks before it is fully ripe. Separate the seed by macerating with water and floating off the pulp. Store in sealed containers at 5°C. Storage longevity is unknown. Sow outdoors in autumn or cold stratify at 1-5°C for 90-115 days if planted in the spring. Sow 2.5 cm deep with shading (Hubbard 1974, Prockter 1976, Young and Young 1986).

References

Elias, T.S. 1980. *The Complete Trees of North America Field Guide and Natural History.* New York: Van Nostrand Reinhold Company. 948p.

Habeck, R.J. 1992. *Rhamnus purshiana. In*: Fischer, William C. (comp.) The Fire Effects Information System [Monograph Online]. Missoula, MT: USDA Forest Service, Intermountain Fire Sciences Laboratory. http://www.fs.fed.us/database/feis/plants/Tree/RHAPUR. Accessed March 4, 1997.

Hubbard, R.L. 1974. *Rhamnus L. Buckthorn.* pp. 704-8 *In*: Schopmeyer, C.S. (tech. coord.) 1974. *Seeds of the Woody Plants in the United States.* Agric. Handbook 450. Washington, DC: USDA Forest Service. 883p.

Pojar, J., and A. MacKinnon. 1994. *Plants of the Pacific Northwest Coast: Washington, Oregon, British Columbia, and Alaska*. Vancouver, BC, Canada: British Columbia Ministry of Forests and Lone Pine Publishing. 527p.

Prockter, N.J. 1976. *Simple Propagation: Propagating by Seed, Division, Layering, Cuttings, Budding and Grafting*. London: Faber and Faber. 246p.

Young, J.A., and C.G. Young. 1986. *Collecting, Processing, and Germinating Seeds of Wildland Plants*. Portland, OR: Timber Press. 236p.

Seeds per kilogram: ~11,020-41,890 (Hubbard 1974)

Vegetative: Hardwood cuttings can be taken in September and October. Plants can also be propagated by layering in early spring (Prockter 1976).

Rhododendron albiflorum
White-flowered rhododendron

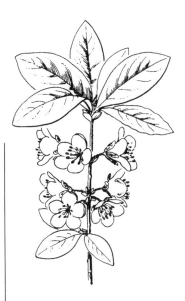

Description
White-flowered rhododendron is an erect, deciduous shrub growing up to 3 m in height. The bark peels and young twigs are covered with reddish hairs. The leaves are alternate but found in clusters along the branch, oblong to elliptic in shape, 4-9 cm long, yellowish-green turning red and orange in the fall, with fine hairs on the surface. The flowers are white, 1-2 cm across, borne in clusters of two to four on the previous year's growth. The fruit is a dry, oval-shaped capsule 6-8 mm long. White-flowered rhododendron is generally not eaten by livestock or wildlife (Haeussler et al. 1990, Pojar and MacKinnon 1994, Taylor and Douglas 1995).

Habitat and Geographic Range
White-flowered rhododendron ranges from British Columbia south in the Olympic and Cascade Mountains to Oregon, east to western Montana and Colorado in the Rocky Mountains. It can be found growing in moist forests and meadows, near streams, and around tree clumps in the subalpine zone from 250 m in the northern part of its range to 1800 m throughout its range. It grows on colluvial or morainal deposits and can frequently be found on shallow bedrock (Haeussler et al. 1990, Hendrickson 1993, Taylor and Douglas 1995).

Propagation
Seed: Flowers appear during late July to early August. There is no information on seed germination specific to white-flowered rhododendron, but most *Rhododendron* spp. do not require prechilling or scarification and only require light for germination (Haeussler et al. 1990).
Vegetative: White-flowered rhododendron reproduces from rhizomes and layering (Haeussler et al. 1990).

References
Haeussler, S., D. Coates, and J. Mather. 1990. Autecology of common plants in British Columbia: A Literature Review. British Columbia Ministry of Forests. FRDA Report-158. 272p.

Hendrickson, D. 1993. The untamed western native: *Rhododendron albiflorum*. *Journal of the American Rhododendron Society* 47(1):8.

Pojar, J., and A. MacKinnon. 1994. *Plants of the Pacific Northwest Coast: Washington, Oregon, British Columbia, and Alaska*. Vancouver, BC, Canada: British Columbia Ministry of Forests and Lone Pine Publishing. 527p.

Taylor, R.J,. and G.W. Douglas. 1995. *Mountain Plants of the Pacific Northwest: A Field Guide to Washington, Western British Columbia, and Southeastern Alaska*. Missoula, MT: Mountain Press Publishing Company. 437p.

Rhododendron macrophyllum
Pacific rhododendron

Description

Pacific rhododendron is an evergreen shrub or small tree that reaches 7.5 m in height. The bark is reddish-brown and scaly. Leaves are alternate, leathery, and shiny, ranging from 6-25 cm long and 3.5-6.5 cm wide. The flowers are trumpet shaped, 3.5-4 cm long, white to pink in color, and are found in terminal clusters. The fruit is a capsule, 1.5-2 cm long and reddish brown. Mountain beaver eat the branches (Elias 1980). Pacific rhododendron provides erosion protection on steep watersheds (Crane 1990).

Habitat and Geographic Range

Pacific rhododendron grows on moist, well-drained sites in the coastal mountains from British Columbia to central California below 1250 m in elevation. It prefers shaded forests and is found in association with coastal redwood, Douglas-fir, and yellow pine (Elias 1980).

Propagation

Seed: Pacific rhododendron blooms between April and July. Collect seed in late summer and fall as soon as the fruit starts to lose its color. Clean the fruit by rubbing or beating it to remove the seed. Seed can be stored at 20°C for up to two years if dried and sealed in glass bottles or polyethylene bags. Mix stored or fresh seed with a small amount of fungicide and sow on a peat:perlite medium that has a thin layer of grit (perlite or pumice) on the surface. Covering with glass or plastic may hasten germination. Seeds should germinate in four to five weeks. Good drainage is crucial (Wilson 1996). Another seeding method is to sow on screened coarse peat moss and water in with a dilute fungicide solution. Then, place uncovered on a bottom-heated bench under intermittent mist and apply fungicide weekly (Wilson 1996). Seeds require light for germination (Crane 1990). Once established, transplant to seedling flats, then to a shade house, and finally to beds for one to two years before outplanting (Olson 1974).
Seeds per kilogram: ~4,409,170 (Olson 1974)

References

Crane, M.F. 1990. *Rhododendron macrophyllum. In*: Fischer, William C. (comp.) The Fire Effects Information System [Monograph Online]. Missoula, MT: USDA Forest Service, Intermountain Fire Sciences Laboratory. http://www.fs.fed.us/database/feis/plants/Tree/RHOMAC. Accessed April 4, 1997.

Elias, T.S. 1980. *The Complete Trees of North America Field Guide and Natural History.* New York: Van Nostrand Reinhold Company. 948p.

Gambrill, K.W. 1978. Rhododendron species propagation and experiences related to dormancy. Combined Proceedings International Plant Propagators Society 28:123-28.

Vegetative: Take stem cuttings from current season's growth from May to September. Soak for a minimum of five minutes in a benomyl solution. Cuttings should be 3.8-7.6 cm long; wound the lower 1-2 cm of the stem to expose two lines of cambium tissue. Dip cuttings in a rooting hormone preparation with a 0.1-1.6% IBA content, then stick in a peat:perlite (1:1) medium and place in a greenhouse with bottom heat (just above 21°C) and mist. When a 3.5-5 cm diameter root ball has formed, transplant to a sawdust:peat (1:1) medium and fertilize. The greenhouse air temperature should be kept at 14°C through the winter. Early the following summer, transfer cuttings outdoors to harden off (Gambrill 1978). Pacific rhododendron can also be propagated by layering (Crane 1990). If aboveground portions are killed, Pacific rhododendron regenerates naturally by sprouting from stem bases and from the root crown (Crane 1990).

Olson, D.F., Jr. 1974. *Rhododendron* L. Rhododendron. pp. 692-703 *In*: Schopmeyer, C.S. (tech. coord.) 1974. *Seeds of the Woody Plants in the United States*. Agric. Handbook 450. Washington, DC: USDA Forest Service. 883p.

Wilson, M.G. 1996. Restoration ecologist, Portland, OR. Personal communication.

Ribes cereum
Squaw currant

References

Helliwell, R. 1987. Forest Plants of the Warm Springs Indian Reservation. Warm Springs, OR: Confederated Tribes of the Warm Springs. 177p.

Link, E. 1993. (ed.) Native Plant Propagation Techniques for National Parks: Interim Guide. East Lansing, MI: Rose Lake Plant Materials Center. 240p.

Pfister, R.D. 1974. *Ribes* L. Currant, gooseberry. pp. 720-27 *In*: Schopmeyer, C.S. (tech. coord.) 1974. *Seeds of the Woody Plants in the United States.* Agric. Handbook 450. Washington, DC: USDA Forest Service. 883p.

Winkler, G., and K.A. Marshall. 1995. *Ribes cereum. In*: Fischer, William C. (comp.) The Fire Effects Information System [Monograph Online]. Missoula, MT: USDA Forest Service, Intermountain Fire Sciences Laboratory. http://www.fs.fed.us/database/feis/plants/Shrub/RIBCER. Accessed April 4, 1997.

Description

Squaw currant, also called wax currant, is an unarmed, deciduous, erect shrub growing up to 1.5 m in height. The leaves are 1-3 cm wide, 1-3 cm long, and three to five lobed. The flowers are tubular in shape, 8-10 mm long, greenish-white to pink and form drooping clusters. The fruit is a bright red berry, 1.2 cm in diameter, and contains numerous seeds. The twigs are an occasional browse for deer and elk during the winter months, and the berries are eaten by birds and small mammals (Helliwell 1987). Squaw currant can play an important role in secondary succession. Its roots stabilize the soil and its foliage may shelter conifer seedlings. The fruit is used for making jam, jelly, or pie. Some Native American tribes used currants for making pemmican. Squaw currant is an alternate host for white pine blister rust which infests five-needled pines and has therefore been a target of various unsuccessful eradication efforts (Winkler and Marshall 1995).

Habitat and Geographic Range

Squaw currant ranges from central and eastern British Columbia south to the Sierra Nevada, northern Arizona, and northern New Mexico (Winkler and Marshall 1995). It grows on open, dry, rocky ground at elevations from 1500 to 4000 m (Helliwell 1987). This species is shade intolerant.

Propagation

Seed: Squaw currant flowers from April to June and the fruit ripens during August. Pick the seed from the branches soon after ripening (Pfister 1974). Clean by macerating in a blender with water, then straining and spreading out to dry. Store in a cool, dry environment. Seeds require scarification to germinate (Winkler and Marshall 1995) and a cold, moist stratification for 12-18 weeks is needed before planting (Link 1993).
Seeds per kilogram: ~443,120-623,900 (Pfister 1974)
Vegetative: Take hardwood cuttings in June. Prepare with heels and dip into a 0.8% concentration of IBA before sticking in a mist bench (Link 1993).

Ribes lacustre
Black gooseberry

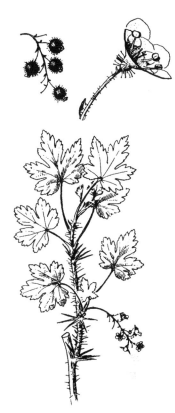

Description
Black gooseberry is a spreading or upright deciduous shrub that reaches 1-1.5 m in height and grows individually or in extensive low thickets. Young stems contain many sharp prickles, while older stems are almost smooth. The leaves are small, five lobed, doubly serrate, and resemble a maple leaf. The drooping raceme has five to fifteen flowers. The fruit is a bristly, purple-black berry, 8.5 mm in diameter, with stalked glands (Haeussler et al. 1990). Black gooseberry fruit is eaten by bears, birds, and rodents. Elk and deer eat the foliage (Winkler and Carey 1995).

Habitat and Geographic Range
Black gooseberry occurs throughout Canada and extends south into the Coast and Cascade ranges to northern California, in the Rocky Mountians to central Colorado and northern Utah, in the Great Lakes states, and in the Appalachian Mountains to West Virginia. It is found in riparian areas, alluvial soils, moist wooded areas, and mountain slopes from 900 to 3500 m in elevation. It is moderately shade tolerant but grows vigorously in canopy openings. (Haeussler et al. 1990, Winkler and Carey 1995)

Propagation
Seed: Black gooseberry begins producing seed at three to five years, with a good crop every two to three years. Flowers bloom from May to June and the fruit ripens in August. Seed can be collected by picking or stripping fruit from the branches. If the seed is not extracted immediately, it should be spread out to dry. Soak dried fruit in water prior to cleaning seed by macerating and washing. Decant the pulp and empty seeds. Seeds remain viable for long periods of time when stored in sealed containers at a low moisture content (Pfister 1974). If spring sowing, seed must be cold, moist stratified at 0°C for 120-200 days (Vories 1980). Acid scarification accomplished by a five-minute soak in 2-10% sulfuric acid can improve germination (Winkler and Carey 1995). Seed is usually sown in the fall to a depth of 0.3-0.6 cm in a moist mineral soil supplied with humus

References
Haeussler, S., D. Coates, and J. Mather. 1990. Autecology of common plants in British Columbia: A literature review. British Columbia Ministry of Forests. FRDA Report 158. 272p.

Pfister, R.D. 1974. *Ribes* L. Currant, gooseberry. pp. 720-27 *In*: Schopmeyer, C.S. (tech. coord.) 1974. *Seeds of the Woody Plants in the United States*. Agric. Handbook 450. Washington, DC: USDA Forest Service. 883p.

Strik, B.C. and A.D. Bratsch. 1990. Growing currants and gooseberries in your home garden. Corvallis, OR: Oregon State University extension bulletin. EC 1361.

Vories, K.C. 1980. Growing Colorado Plants from Seed: A State of the Art. Vol. 1: Shrubs. Ogden, UT: Intermountain Forest and Range Experiment Station. USDA Forest Service Gen. Tech. Rep. INT-103. 80p.

Winkler, G., and J.H. Carey. 1995. *Ribes lacustre. In*: Fischer, William C. (comp.) The Fire Effects Information System [Monograph Online]. Missoula, MT: USDA Forest Service, Intermountain Fire Sciences Laboratory. http://www.fs.fed.us/database/feis/plants/Shrub/RIBLAC. Accessed April 4, 1997.

and covered with 5-7.5 cm of mulch. Outplant after two years.

Seeds per kilogram: ~1,135,360 (Pfister 1974)

Vegetative: Take hardwood cuttings 15-20 cm long from one-year-old wood in the fall. Make a flat bottom cut just below a bud and a slanted top cut about 1.3 cm above. Stick the cuttings in well-drained soil with only one or two buds extending from the soil. Plant rooted cuttings in a permanent location after one or two years' growth (Strik and Bratsch 1990). Black gooseberry also regenerates via layering (Winkler and Carey 1995).

Rosa gymnocarpa
Baldhip rose

Description

Baldhip rose is a slim-stemmed, erect, deciduous shrub with many slender, straight prickles that reaches 1.5 m in height. The leaves are alternate, pinnately compound with five to nine leaflets, 1-4 cm long, doubly serrate, with a pair of wedge-shaped stipules at the base of the rachis. The small pink flowers are usually borne singly. The fruit is a red-orange rose hip and the calyx soon dehisces. The hips are eaten by birds and small mammals (Helliwell 1987).

Habitat and Geographic Range

Baldhip rose can be found widely from low to high elevations and ranges from southern British Columbia east to northwest Montana, south to southern Idaho, and into central California (Helliwell 1987).

Propagation:

Seed: Hand pick the hips in August and September after the green color turns reddish. Clean the seed by macerating the fruit in water and recovering the seed by flotation. Seed can be stored dry for two to four years in sealed containers at 1-3°C. Stored seed must be cold stratified at 4°C for 90 days before planting. Germination is greatest if seeds are sown immediately after cleaning. Cover seed with a shallow layer of soil and mulch with sawdust or oat straw (Gill and Pogge 1974).
Seeds per kilogram: ~61,730 (Gill and Pogge 1974)
Vegetative: Buis (1996) reports poor success with hardwood cuttings and recommends semi-hardwood cuttings. Cuttings should contain three to four nodes each. Strip the basal end leaves and treat the cutting with a medium-strength rooting hormone, stick into a perlite:vermiculite (1:1) mixture and set on a misting bench with light mist at 21°C. Once rooted, move to a hoop house or shade house.

References

Buis, S. 1996. Owner, Sound Native Plants, Olympia, WA. Personal communication.

Gill, J.D., and F.L. Pogge. 1974. *Rosa* L. Rose. pp. 732-37 *In*: Schopmeyer, C.S. (tech. coord.) 1974. *Seeds of the Woody Plants in the United States*. Agric. Handbook 450. Washington, DC: USDA Forest Service. 883p.

Helliwell, R. 1987. Forest Plants of the Warm Springs Indian Reservation. Warm Springs, OR: Confederated Tribes of the Warm Springs. 177p.

Rosa nutkana
Nutka rose

References

Buis, S. 1996. Owner, Sound Native Plants, Olympia, WA. Personal communication.

Gill, J.D., and F.L. Pogge. 1974. *Rosa* L. Rose. pp. 732-37 *In*: Schopmeyer, C.S. (tech. coord.) 1974. *Seeds of the Woody Plants in the United States*. Agric. Handbook 450. Washington, DC: USDA Forest Service. 883p.

Helliwell, R. 1987. Forest Plants of the Warm Springs Indian Reservation. Warm Springs, OR: Confederated Tribes of the Warm Springs. 177p.

King County Department of Public Works, Surface Water Management Division. 1994. Northwest Native Plants, Identification and Propagation for revegetation and restoration projects. King County, WA. 68p.

Description

Nutka rose is a deciduous, perennial shrub with erect or trailing stems that reach 1-1.8 m (Reed 1993). The stems have large, curved prickles at the nodes (Helliwell 1987). The leaves are compound with five to seven leaflets (Reed 1993). The leaflets are oval or elliptical, 2.5-3.8 cm long, and serrated from the tip to just past the midpoint. The flowers are 5 cm across, borne singularly, and pink to purple in color (King County 1994). The fruit is a pear-shaped hip, 12-20 mm long, with a persistent calyx (Helliwell 1987, King County 1994). Nutka rose has excellent soil-binding characteristics and is a good source of food and cover for many wildlife species (Reed 1993, King County 1994).

Habitat and Geographic Range

Nutka rose ranges from Alaska south to California and east to New Mexico and western Montana. It can be found in floodplains, open streambanks, and meadows from low to mid- elevations. It grows on nitrogen-rich, fresh, moist, well-drained clayey-loam, sandy-loam, or sandy soils in moderately dry to moist climates. Nutka rose is both sun and shade tolerant but produces more fruit as its exposure to light increases (Reed 1993, Pojar and MacKinnon 1994).

Propagation:

Seed: Nutka rose reaches sexual maturity at two to five years of age with good seed crops every other year. It flowers from May through June and the fruit ripens in early fall and remains on the plant through winter (Reed 1993). Collect the fruit from August through September and either dry and crush or soak in water and macerate to remove the seed (King County 1994). The seeds require a period of after-ripening. If spring sown, a warm stratification followed by a cold stratification is necessary for germination, or seed can be sown fresh in the fall. Sow into propagating flats in a finely milled peat:vermiculite growing medium (Buis 1996).
Seeds per kilogram: ~66,135-132,275 (Gill and Pogge 1974)

Vegetative: Nutka rose sprouts from the root crown and also grows from small offshoots from the parent plant (Reed 1993, King County 1994).

Pojar, J., and A. MacKinnon. 1994. *Plants of the Pacific Northwest Coast: Washington, Oregon, British Columbia, and Alaska.* Vancouver, BC, Canada: British Columbia Ministry of Forests and Lone Pine Publishing. 527p.

Reed, W.R. 1993. *Rosa nutkana. In*: Fischer, William C. (comp.) The Fire Effects Information System [Monograph Online]. Missoula, MT: USDA Forest Service, Intermountain Fire Sciences Laboratory. http://www.fs.fed.us/database/feis/plants/Shrub/ROSNUT. Accessed September 11, 1996.

Rosa pisocarpa
Cluster rose

References

Buis, S. 1996. Owner, Sound Native Plants, Olympia, WA. Personal communication.

Gilkey, H.M., and L.J. Dennis. 1980. *Handbook of Northwestern Plants.* Corvallis, OR: Oregon State University Bookstores, Inc. 507p.

Peck, M.E. 1961. *A Manual of the Higher Plants of Oregon.* Portland, OR: Binfords & Mort Publishers. 936p.

USDA Forest Service. 1988. *Range Plant Handbook.* New York: Dover Publications, Inc. 816p.

Description

Cluster rose is a slender, erect, shrub reaching 1-4 m in height. The stems are scarcely armed with straight or slightly curved prickles. The leaves are compound, 5-10 cm long, with five to seven leaflets. The leaflets are oval shaped, dark green above and puberulent beneath. The rachis often has a few prickles. The flowers are 3.5-4.5 cm in diameter with the calyx lobes elongated and smooth or glandular and are generally borne several in a corymb, but sometimes found solitary. The petals are broad and deep pink in color with a slight tinge of magenta. The fruit is a hip, globose or broadly ovoid, 1 cm long, with the calyx lobes persistent. Cluster rose is regarded as good sheep browse (Peck 1961, Gilkey and Dennis 1980, USDA 1988).

Habitat and Geographic Range

Cluster rose ranges from British Columbia south to Oregon and Idaho, and possibly into California and Utah. It is found growing scattered along streams, dry ravines, in thickets and open woods mostly west of the Cascade Mountains (Peck 1961, USDA 1988).

Propagation:

Seed: Collect seed in the fall by hand. It is best to soak the seed in water for five to seven days in a warm place before cleaning by macerating in water and floating away the pulp. Seed requires a period of after-ripening, accomplished by warm stratification followed by cold stratification if spring sown. Seed can also be sown in containers in the fall with a standard potting mix and overwintered in a hoop house or outside. If kept outside, they can be mulched (Buis 1996).

Rosa woodsii
Wood's rose

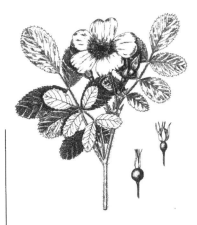

Description

Wood's rose is a fast-growing, long-lived, deciduous shrub. It grows from 0.5-2 m in height. The leaves are alternate and pinnate with five to seven obovate to oblong-obovate leaflets which are 2-6 cm long with sharply serrated margins. There are usually two stout thorns at the base of each leaf and straight to slightly curved prickles 0.5-1.0 cm long on the stems. The five-petaled flowers are a deep pink or occasionally white. The fruit is a stoney achene, 3-4 mm long, within a 6-15 mm wide berrylike hip. Leaves are browsed by big game, livestock, porcupines, and beavers. Dried fruits are eaten by many birds and mammals during the winter. Dense thickets are used as nesting and cover for many birds and small mammals. Wood's rose is a good species to plant for erosion control due to its good survivability and revegetation characteristics (Vories 1981, Snyder 1991, Tesky 1992, Strickler 1993).

Habitat and Geographic Range

Wood's rose ranges from British Columbia to western Ontario and Minnesota, south to Missouri, Nebraska, Arizona, and northern Mexico. Its elevational range is from 1000 to 3500 m. It grows on a wide range of soil types and textures but prefers a moist, well-drained clay loam, sandy loam, or sandy soil. It can be found on bluffs, dry grassy slopes, and sandhills in the prairies, but is more common on riverbanks and clearings in boreal and subalpine forests. Wood's rose is rarely found where the average precipitation is less than 30 cm and does best in moderate shade to full sun (Tesky 1992, Strickler 1993).

Propagation

Seed: *Rosa* spp. take two to five years before they start flowering and producing seed. Flowers bloom from June to July and the fruits ripen from late summer through fall. Hand pick the hips or knock them into hoppers or containers. Clean by macerating the hips in water and floating off the pulp and empty seed. Seed can be stored dry in sealed containers at 1-3°C and will remain viable for two to five years. Sow fresh seed in the fall; spring-

References

Anderson, J. 1996. Sevenoaks Native Nursery, Corvallis, OR. Personal communication.

Gill, J.D., and F.L. Pogge. 1974. *Rosa* L. Rose. pp. 732-37 *In*: Schopmeyer, C.S. (tech. coord.) 1974. *Seeds of the Woody Plants in the United States.* Agric. Handbook 450. Washington, DC: USDA Forest Service. 883p.

Snyder, L.C. 1991. Native Plants for Northern Gardens. Anderson Horticultural Library, University of Minnesota. 277p.

Strickler, D. 1993. *Wayside Wildflowers of the Pacific Northwest.* Columbia Falls, MT: Flower Press. 272p.

Tesky, J.L. 1992. *Rosa woodsii*. *In*: Fischer, William C. (comp.) The Fire Effects Information System [Monograph Online]. Missoula, MT: USDA Forest Service, Intermountain Fire Sciences Laboratory. http://www.fs.fed.us/database/feis/plants/Shrub/ROSWOO. Accessed June 25, 1996.

Vories, K.C. 1980. Growing Colorado Plants from Seed: A State of the Art. Vol. 1: Shrubs. Ogden, UT: Intermountain Forest and Range Experiment Station. USDA Forest Service Gen. Tech. Rep. INT-103, 80p.

sown seed should undergo a period of warm stratification followed by a period of cold stratification at 4°C for three to six months. Recommended planting depth is 0.5-2.0 cm with an oat straw or sawdust mulch (Gill and Pogge 1974, Vories 1981, Snyder 1991).

Seeds per kilogram: ~77,160-143,300 (Gill and Pogge 1974)

Vegetative: Wood's rose spreads vegetatively through underground rhizomes, sprouting from the root crown, and layering (Tesky 1992). While hardwood cuttings have not proven very successful, softwood cuttings 15 cm long can be taken from mid- to late June into early July (Anderson 1996). Propagation can also be accomplished by division (Snyder 1991).

Rubus idaeus
Red raspberry

Description
Red raspberry is a deciduous shrub with erect stems, reaching 1-2 m tall. The stems originate from a perennial, branching rhizome and contain abundant bristles, prickles, and hairs. Flowering lateral branches are produced during the second year of growth. The leaves are alternate and pinnately compound with three to seven irregularly shaped, toothed lobes. Small, showy, perfect white flowers are borne in clusters of one to four in a compound cyme. The fruit is an aggregate of many red or pinkish-purple drupelets. The berry is food for birds, squirrels, chipmunks, mice, and racoons. Deer and rabbit browse the leaves and stems (Haeussler et al. 1990).

Habitat and Geographic Range
Red raspberry ranges from Alaska to Newfoundland, south to North Carolina and Tennessee, and in the West to Arizona, California, and northern Mexico. It is a shade-intolerant species and can be found on disturbed sites in areas ranging from valley bottoms and stream banks to clearings in forests and subalpine elevations. It grows on imperfectly to well-drained sandy loams to silty clay loams, but grows best on moderately well-drained soils (Haeussler et al. 1990, Tirmenstein 1990).

Propagation
Seed: Most species of raspberry produce good seed crops nearly every year but seed production can vary annually depending on climatic factors and the age of the stems. The fruit of red raspberry ripens from late June to October and can be hand picked. Extract the seeds by macerating in water and floating off the pulp and empty seed. Dried seed can be stored at 5°C for several years (Brinkman 1974). Germination is best when seed is scarified and sown in the fall. Spring-sown seeds require a warm stratification at 20-30°C for 90 days, followed by a cold stratification at 2-5°C for an additional 90 days (Tirmenstein 1990). Germination is improved when seeds are scarified in either sulphuric acid for 20-60 minutes or a 1% solution of sodium hyperchlorite for seven days. Sow in drills and lightly

References
Brinkman, K.A. 1974. *Rubus* L. Blackberry, raspberry. pp. 738-43 *In*: Schopmeyer, C.S. (tech. coord.) 1974. *Seeds of the Woody Plants in the United States*. Agric. Handbook 450. Washington, DC: USDA Forest Service. 883p.

Haeussler, S., D. Coates, and J. Mather. 1990. Autecology of common plants in British Columbia: A literature review. British Columbia Ministry of Forests. 272p.

Miller, P. 1993. Propagation of red raspberries. Combined Proceedings International Plant Propagators Society 43:301.

Tirmenstein, D. 1990. *Rubus idaeus*. *In*: Fischer, William C. (comp.) The Fire Effects Information System [Monograph Online]. Missoula, MT: USDA Forest Service, Intermountain Fire Sciences Laboratory. http://www.fs.fed.us/database/feis/plants/Shrub/RUBIDA. Accessed June 25, 1996.

cover with soil. Sowing seeds at greater depths with subsequent exposure to light can produce better germination than shallow plantings, possibly due to greater soil moisture (Tirmenstein 1990). Mulching is suggested over winter (Brinkman 1974).

Seeds per kilogram: ~667,985-846,560 (Brinkman 1974)

Vegetative: Natural vegetative regeneration occurs through stolons, rhizomes, and basal stem buds. Dense raspberry thickets form from the roots or stems of parent plants which separate to form individual plants with the deterioration of connecting tissue. The adventitious shoots produced during the growing season elongate, pushing up through the soil and root. When dormant, the new plants can be dug up and transplanted or held in cold storage until outplanting. Root cuttings can be taken when the plant is dormant if care is taken to prevent drying of the roots before planting (Miller 1993).

Rubus lasiococcus
Dwarf bramble

Description
Dwarf bramble is a trailing perennial shrub, approximately 10 cm tall, with slender, unarmed stems that grow up to 2 m long. The strawberry-like leaves are alternate, deciduous with some persisting over winter, three lobed and 3-6 cm wide. The flowers are white, one or two per stalk, with 5-10 mm long petals, and contain many stamens and pistils. The fruit is a bright red aggregation of druplets with many hairs (Helliwell 1987, Pojar and MacKinnon 1994). The berries are eaten by a number of mammals.

Habitat and Geographic Range
Dwarf bramble grows in moist, shaded woods and open places such as thickets and logged areas. It can be found in the Cascade Mountains north to British Columbia at mid- to high elevations (Peck 1961, Helliwell 1987, Pojar and MacKinnon 1994).

Propagation
Vegetative: Cuttings can be taken from runners during spring to mid-summer. Each cutting should have at least two nodes. Dip the basal end into a light-strength powered rooting hormone and stick into a perlite:vermiculite (1:1) mixture. Set on a mist bench at 21°C and do not allow to get too wet. Once the cuttings have established roots, transplant to a regular potting soil and move to a greenhouse and place in a heavily shaded area (Buis 1996).

References
Buis, S. 1996. Owner, Sound Native Plants, Olympia, WA. Personal communication.

Helliwell, R. 1987. Forest Plants of the Warm Springs Indian Reservation. Warm Springs, OR: Confederated Tribes of the Warm Springs. 177p.

Peck, M.E. 1961. *A Manual of the Higher Plants of Oregon.* Portland, OR: Binfords & Mort Publishers. 936p.

Pojar, J., and A. MacKinnon. 1994. *Plants of the Pacific Northwest Coast: Washington, Oregon, British Columbia, and Alaska.* Vancouver, BC, Canada: British Columbia Ministry of Forests and Lone Pine Publishing. 527p.

Rubus parviflorus
Thimbleberry

References

Buis, S. 1996. Owner, Sound Native Plants, Olympia, WA. Personal communication.

Haeussler, S., D. Coates, and J. Mather. 1990. Autecology of common plants in British Columbia: A literature review. British Columbia Ministry of Forests. FRDA Report-158. 272p.

Link, E. 1993. (ed.) Native Plant Propagation Techniques for National Parks: Interim Guide. East Lansing, MI: Rose Lake Plant Materials Center. 240p.

Pojar, J., and A. MacKinnon. 1994. *Plants of the Pacific Northwest Coast: Washington, Oregon, British Columbia, and Alaska.* Vancouver, BC, Canada: British Columbia Ministry of Forests and Lone Pine Publishing. 527p.

USDA Forest Service. 1988. *Range Plant Handbook.* New York: Dover Publications, Inc. 816p.

Description

Thimbleberry is a deciduous, rhizomatous shrub growing up to 2.5 m tall. This species sprouts short-lived, erect, stems with no prickles. The maple-shaped leaves are alternate, grow up to 25 cm and have three to seven toothed lobes. They have a soft, crinkly surface and are finely pubescent on both sides. The flowers are borne three to eleven in a long-stemmed terminal cluster, up to 4 cm across, with white crinkled petals. The fruit is a shallowly domed aggregate of small, red, hairy drupelets. The leaves, stem, and berries are browse for the black-tailed deer, and the berries are eaten by many different birds (Haeussler et al. 1990, Pojar and MacKinnon 1994).

Habitat and Geographic Range

Thimbleberry ranges from Alaska to California, east to New Mexico, South Dakota, Michigan, and into western Ontario. It can be found at elevations up to 2700 m, but is more common up to 2100 m in the Pacific Northwest. It grows in moist, shaded places such as along streams, in moist draws, avalanche tracks, and open woods, and on wooded hillsides. It prefers a sandy loam soil that is rich in humus, but also grows on rocky sites with thin soils (USDA 1988, Pojar and MacKinnon 1994).

Propagation

Seed: Collect seed before or as soon as the berries are ripe. Place them into a container of water to soak for a few days before macerating. Seed requires no pretreatment. Store damp in a refrigerator over the winter, sow in Feburary in a standard potting mix, and place in a hoop house (Buis 1996).

Seeds per kilogram: ~449,735 (Link 1993)

Vegetative: Thimbleberry can be propagated by cuttings and rhizomes.

Rubus spectabilis
Salmonberry

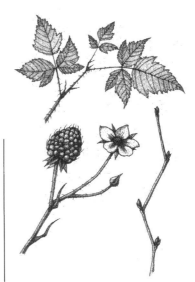

Description
Salmonberry is a deciduous shrub, 0.5-3 m tall, with weak, sparsely thorned stems and light brown exfoliating bark. The leaves are pinnately compound, with three leaflets which have doubly serrate margins and are oval shaped with pointed tips. The flowers are borne singly or in clusters of two or three and are pink to dark red in color. The aggregate fruit is composed of drupelets ranging in color from salmon to dark red. Birds, small mammals, and bears eat the fruit. Although sometimes considered a nuisance by land managers, it does provide important wildlife habitat and helps stabilize the soil (Haeussler et al. 1990, King County 1994).

Habitat and Geographic Range
Salmonberry grows on wet slopes or valleys, along streambanks, and in ravines. It grows from southeast Alaska to the Santa Cruz mountains of California, and from the Pacific Ocean east into Idaho and Montana. It is a moderately shade-tolerant understory species and can often be found under red alder in forested wetlands. It grows well on disturbed sites and is good at preventing soil erosion (King County 1994, Jensen et al. 1995).

Propagation
Seed: Fruit ripens from June through August and can be collected by hand. Extract seeds by macerating in water and floating off the pulp and empty seed. Seeds should be dried for storage, and will keep for several years at 5°C. A warm stratification of 20-30°C for 90 days followed by a cold stratification period of 2-5°C is necessary for spring-sown seeds, although fall sowing provides best germination. Germination is improved if seeds are scarified with sulphuric acid for 20-60 minutes or with a 1% solution of sodium hyperchlorite for seven days prior to cold stratification. Sow in drills and lightly cover with soil and mulch over winter (Brinkman 1974).
Seeds per kilogram: ~315,255 (Brinkman 1974)

References
Albright, M. 1996. Greenhouse Manager. USDI National Park Service, Olympic National Park, Port Angeles, WA. Personal communication.

Brinkman, K.A. 1974. *Rubus* L. Blackberry, raspberry. pp. 738-43 *In*: Schopmeyer, C.S. (tech. coord.) 1974. *Seeds of the Woody Plants in the United States*. Agric. Handbook 450. Washington, DC: USDA Forest Service. 883p.

Haeussler, S., D. Coates, and J. Mather. 1990. Autecology of common plants in British Columbia: A literature review. British Columbia Ministry of Forests. FRDA Report 158. 272p.

Jensen, E.C., D.J. Anderson, J.C. Zasada, and J.C. Tappeiner II. 1995. The reproductive ecology of broadleaved trees and shrubs: salmonberry, *Rubus Spectabilis* Pursh. Corvallis, OR: Forest Research Laboratory, Oregon State University. Research Publication 9e. 7p.

King County Department of Public Works, Surface Water Management Division. 1994. Northwest Native Plants, Identification and Propagation for Revegetation and Restoration Projects. King County, WA.

Vegetative: Salmonberry plants reproduce readily by layering, basal sprouting, and rhizomes (Jensen et al. 1995). Small offshoots growing from the parent plants are easily transplanted. Pull the rooted tips of larger plants and plant into one-gallon pots. Hardwood cuttings can also be taken (King County 1994, Albright 1996); they should be 1-2.5 cm in diameter and 45 cm or more in length with at least three nodes. Rooting invariably occurs at the base of a cutting and at nodes with leaf buds. Store hardwood cuttings over winter in damp sawdust or peat moss; this promotes callusing and prevents desiccation. As with hardwood cuttings of other species, vigorous rooting can be enhanced in *Rubus* spp. by using a liquid rooting hormone and burying the cuttings in damp wood shavings (Albright 1996).

Rubus ursinus
Trailing blackberry

Description

Trailing blackberry is a low-growing, trailing or climbing evergreen shrub that can grow up to 6 m long and build moundlike thickets. The stems are round, green to reddish-brown, glaucous, and armed with slender, small based, straight or recurved prickles. The leaves are pinnately compound with three to five leaflets and doubly serrate margins. The flowers are white, dioecious, and are borne in clusters of two to fifteen near the ends of the branches. The fruit is a black, oblong or conical-shaped aggregate of drupelets about 2.5 cm long. Birds and small mammals eat the fruit. Trailing blackberry grows well on poor soils, which makes it an excellent species for erosion control (Helliwell 1987, Tirmenstein 1989, Randall et al. 1994).

Habitat and Geographic Range

Trailing blackberry ranges from British Columbia to northern California and east to central Idaho. It also grows through southern California into Mexico. Its elevational range is from sea level along the Pacific Coast to middle elevations farther inland. Trailing blackberry grows well on a variety of unproductive infertile soils and its main requirement for vigorous growth is adequate soil moisture. It tolerates a wide range of soil texture and pH and can be found growing in warm, open areas, dense woodlands, prairies, disturbed sites, clearings, and canyons (Tirmenstein 1989).

Propagation

Seed: Most *Rubus* species produce good seed crops nearly every year. Trailing blackberries flower from April to June. Fruit ripens approximately two months after flowering and seed is dispersed approximately one month after that. The seed has a hard, impermeable coat and a dormant embryo, making germination slow. Most species of blackberry require a warm stratification of 20-30°C for 90 days followed by a cold stratification of 2-5°C for another 90 days. A sulfuric acid treatment prior to cold stratification may also increase germination. Plant scarified seed in late summer or early fall. Plant

References

Brinkman, K.A. 1974. *Rubus* L. Blackberry, raspberry. pp. 738-43 *In*: Schopmeyer, C.S. (tech. coord.) 1974. *Seeds of the Woody Plants in the United States*. Agric. Handbook 450. Washington, DC: USDA Forest Service. 883p.

Conrad, C.E. 1987. Common Shrubs of Chaparral and Associated Ecosystems of Southern California. USDA Forest Service. Gen. Tech. Rep. PSW-99. 86p.

Helliwell, R. 1987. Forest Plants of the Warm Springs Indian Reservation. Warm Springs, OR: Confederated Tribes of the Warm Springs. 177p.

Randall, W.R., R.F. Keniston, D.N. Bever, and E.C. Jensen. 1994. *Manual of Oregon Trees and Shrubs*. Corvallis, OR: Oregon State University Bookstores. 305p.

Tirmenstein, D.A. 1989. *Rubus ursinus*. *In*: Fischer, William C. (comp.) The Fire Effects Information System [Monograph Online]. Missoula, MT: USDA Forest Service, Intermountain Fire Sciences Laboratory. http://www.fs.fed.us/database/feis/plants/Shrub/RUBURS. Accessed January 24, 1997.

seed that has been stratified and scarified in the spring. Cover with 3-5 mm of soil (Brinkman 1974, Tirmenstein 1989).

Seeds per kilogram: ~846,560 (Brinkman 1974)

Vegetative: Trailing blackberry can be an aggressive, vigorous spreading plant. It sprouts readily from suckers or nonrhizomatous sprouts or from trailing stems which root at the nodes (Conrad 1987, Tirmenstein 1989).

Sambucus cerulea
Blue elderberry

Description

Blue elderberry is a deciduous shrub or small tree that grows up to 15 m tall. The bark is irregularly furrowed and dark reddish brown in color. The leaves are opposite, pinnately compound, with five to nine leaflets which are narrowly oval shaped and coarsely toothed at the margins. The white flowers are produced in a broad cyme at the end of the branchlets. The fruit is berrylike, 6-10 mm in diameter, covered in a waxlike coating, and is dark blue to black in color. The fruit is eaten by songbirds, rabbits, and squirrels. Small mammals eat the fruit and bark, while deer and elk browse the twigs and foliage. The stems are used to make flutes, and the berries are used in making pies, jellies, and wine. The wood is sometimes used for fence posts (Elias 1980).

Habitat and Geographic Range

Blue elderberry grows in dry to moist, well-drained sites and is most commonly found along streambanks, edges of fields, and in rocky pastures. It can be found from sea level to 2700 m in elevation and ranges from southern British Columbia to southern California (Elias 1980).

Propagation

Seed: The fruit ripens in August and September; collect by stripping or cutting the clusters from the branches. Either dry the fruit, run it through a macerator with water and float off the pulp and empty seeds, or crush the fruit, dry, and use the seed with no additional cleaning. Seed can be stored dry for several years at 5°C. Sow in the fall soon after collection, or stratify and sow in the spring. Blue elderberry seeds have dormant embryos and require a warm stratification for 60-90 days followed by a cold stratification period of 90-100 days at 5°C. Sow to a depth of 0.6 cm and cover with a thin layer of sawdust mulch. Mulch should be used on fall-sown seedbeds. Seedlings are usually large enough for outplanting after one year (Brinkman 1974).
Seeds per kilogram: ~257,935-570,990 (Brinkman 1974)
Vegetative: Blue elderberry can be propagated by hardwood cuttings from the previous season's growth. Take cuttings during winter and with a heel so as not to expose the pith (Huxley et al. 1992).

References

Brinkman, K.A. 1974. *Sambucus* L. Elder. pp. 754-59 *In*: Schopmeyer, C.S. (tech. coord.) 1974. *Seeds of the Woody Plants in the United States.* Agric. Handbook 450. Washington, DC: USDA Forest Service. 883p.

Elias, T.S. 1980. *The Complete Trees of North America Field Guide and Natural History.* New York: Van Nostrand Reinhold Company. 948p.

Huxley, A., M. Griffiths, and M. Levy. 1992. *New Royal Horticultural Society Dictionary of Gardening.* Vol 4. London: Macmillan Press Ltd. 189p

Sambucus racemosa
Red elderberry

Description

Red elderberry is a deciduous shrub that grows up to 6 m in height. The leaves are 15-30 cm long, opposite and compound consisting of five to seven oblong to ovate leaflets, pointed at the tip, with a fine-toothed margin. The flowers are found in tiny, white, conical clusters, and ripen into bright red berries in the summer. It has good soil-binding characteristics and the fruit is eaten by many birds and small mammals (King County 1994).

Habitat and Geographic Range

Red elderberry is found across North America from Newfoundland to Alaska. It is restricted to moist, cool sites in the south and extends southward into California in the coastal mountains, into Arizona and New Mexico in the Rocky Mountains, and into Georgia and Tennessee in the Appalachian highlands. Its elevational range is 1000-3500 m (Crane 1989).

Propagation

Seed: Red elderberry flowers from April to July and fruit ripens from June through September. Collect by either stripping or cutting the clusters from the branches. Prepare the seed for planting by drying the fruit, macerating it with water and floating off the pulp and empty seeds, or crushing and drying it for use without any additional cleaning. Seeds can be stored dry for several years at 5°C. Sow in the fall shortly after collection or stratify and sow in the spring, but germination is frequently not complete until the second spring. Pregermination treatment consists of a five-minute sulphuric acid treatment followed by a two-day soak in water, then a warm and cold stratification. Sow to a depth of 0.6 cm and cover with a thin sawdust mulch layer. Seedlings are usually large enough for outplanting after one year (Brinkman 1974). Another source (Wilson 1996) reports that seed can be sown in the spring without stratification other than being cold stored for a period of time. For this method, plant the seed on a peat:sand:perlite media by using a salt shaker and sprinkling over the surface. Then sprinkle with a fine covering of perlite. Place a sheet of glass over the seeded

References

Aubin, L. 1982. The propagation of *Sambucus racemosa* 'Sutherland golden' by softwood cuttings. *The Plant Propagator*. International Plant Propagators Society 28(1): 13-14.

Brinkman, K.A. 1974. *Sambucus* L. Elder. pp. 754-59 *In*: Schopmeyer, C.S. (tech. coord.) 1974. *Seeds of the Woody Plants in the United States*. Agric. Handbook 450. Washington, DC: USDA Forest Service. 883p.

tray to maintain high humidity. After three weeks, the seed will begin to germinate. Leave the glass in place but lift it 1 cm to provide some air circulation and help prevent damping-off fungi. When germination is complete, remove the glass and keep the seedlings watered and fertilized until late winter. Then separate them and plant into individual pots and allow to grow another year before outplanting.

Seeds per kilogram: ~423,280-831,130 (Brinkman 1974)

Vegetative: Take softwood cuttings in June, as soon as the new wood is sturdy. The length of the cuttings depends on the length of the internodes. Make a basal cut just below a node and remove 30-40% of the leaves. Hardwood and semi-hardwood cuttings root very well also. Cuttings can be stored over winter in damp sawdust or peat moss. In addition to an IBA rooting hormone, cuttings can be dipped in a Captan® solution for a few minutes. In late February to early May, stick cuttings in a 1:1 peat:perlite mixture or potting soil in containers with ample room for root growth. Cuttings root within 14-21 days if kept under hot and humid conditions, but too much water will cause the leaves to rot. Cuttings do not do well in cold storage and should be planted in the field with enough time to be established before winter (Aubin 1981). Red huckleberry can also regenerate via sprouts, rhizomes, and layering (Crane 1989).

Crane, M.F. 1989. *Sambucus racemosa ssp. pubens. In*: Fischer, William C. (comp.) The Fire Effects Information System [Monograph Online]. Missoula, MT: USDA Forest Service, Intermountain Fire Sciences Laboratory. http://www.fs.fed.us/database/feis/plants/Tree/SAMRACP. Accessed April 4, 1997.

King County Department of Public Works, Surface Water Management Division. 1994. Northwest Native Plants, Identification and Propagation for Revegetation and Restoration projects. King County, WA.

Wilson, M.G. 1996. Restoration ecologist, Portland, OR. Personal communication.

Shepherdia canadensis
Russet buffaloberry

References

Hellyer, A.G.L. 1972. *Sanders' Encyclopaedia of Gardening*. London: Collengridge Books. 534p.

Kruckeberg, A.R. 1982. *Gardening with Native Plants of the Pacific Northwest*. Seattle, WA: University of Washington Press. 252p.

Link, E. 1993. (ed.) Native Plant Propagation Techniques for National Parks: Interim Guide. East Lansing, MI: Rose Lake Plant Materials Center. 240p.

Randall, W.R., R.F. Keniston, D.N. Bever, and E.C. Jensen. 1994. *Manual of Oregon Trees and Shrubs*. Corvallis, OR: Oregon State University Bookstores. 305p.

Thilenius, J.F., K.E. Evans, and E.C. Garrett. 1974. *Shepherdia* Nutt. Buffaloberry. pp. 771-73 *In*: Schopmeyer, C.S. (tech. coord.) 1974. *Seeds of the Woody Plants in the United States*. Agric. Handbook 450. Washington, DC: USDA Forest Service. 883p.

Description

Russet buffaloberry, also called Canadian buffaloberry or nannyberry, is a deciduous, thornless shrub growing 1-3 m high. The leaves are ovate to obovate with entire margins. The small, yellow, male and female flowers are found on separate plants growing either singly or in clusters on the branchlets. The fruit is an ovoid, drupelike achene. The plant is occasionally browsed by deer. The berries provide food for bears and birds. Buffaloberry is a nitrogen fixer and is a good plant for restoration projects (Thilenius et al. 1974).

Habitat and Geographic Range

Russet buffaloberry grows on dry or moist well-drained sites at elevations of 1500 to 2500 m. Its range is from Alaska to east-central Oregon, east to New England, and the Appalachian Mountains (Randall et al. 1994).

Propagation

Seed: Russet buffaloberry flowers from April to June and the fruit ripens from June to August, when it turns yellow or red. Collect the fruit by stripping or flailing from the bush onto a canvas tarp. Clean it by running it through a macerator with water and floating the pulp off. Keep the cleaned seed dry. Germination of russet buffaloberry seed is increased by scarification with sulfuric acid, cold stratification for 60 days and diurnally alternating temperatures of 20°C and 30°C. Cover the seed with 0.6 cm of soil and 1.3-2.5 cm of straw mulch. Outplanting can be done with two-year-old stock (Thilenius et al. 1974).

Seeds per kilogram: ~114,638 (Link 1993)

Vegetative: Propagation of russet buffaloberry is best from cuttings. Stick root cuttings in February or March in ordinary soil outdoors (Kruckeberg 1982). Layering of shoots can be done in autumn (Hellyer 1972).

Sorbus sitchensis
Sitka mountain-ash

Description
Sitka mountain-ash is a large, deciduous shrub, or occasionally a small tree, growing up to 9 m in height with a round, open crown. The bark is light gray, smooth and scaling with age. The leaves are alternate, pinnately compound, with seven to eleven lanceolate, sharply serrate leaflets. White flowers are borne in terminal corymbs with fifteen to sixty flowers per head. The bright red fruit is a round berry, 0.6-1.2 cm in diameter, found in clusters at the end of the branchlet, and contains one or two ovate-shaped seeds. The berries provide food for many birds and small mammals. The twigs are browsed by deer, moose, and elk, and the berries, leaves, and stems are eaten by bears. Sitka mountain-ash can be used for streambank rehabilitation but may produce allelopathic substances that inhibit the growth of Douglas-fir seedlings (Elias 1980, Matthews 1993).

Habitat and Geographic Range
Sitka mountain-ash grows on moist, rich, well-drained soils along stream borders or rocky hillsides, and is usually found among conifers. It can be found at elevations ranging from 760 to 3050 m from Alaska to central California, and east into northern Idaho and northwestern Montana (Elias 1980, Matthews 1993).

Propagation
Seed: Sitka mountain-ash begins producing seed at approximately fifteen years of age and usually produces a good crop every year. Fruit ripens from August to October and should be picked or shaken from the tree soon after to prevent losses to birds. Extract seed using a macerator and remove pulp by flotation, screening, or skimming. Dry and clean seed for storage. Seed stored in sealed containers at 6-8% moisture content and 1-3°C can be kept for up to eight years with little loss of viability. Sow unstratified seed in the fall or early winter. Spring-sown seed requires a cold stratification of 90-140 days at 1-5°C in moist peat or soil. Sow cleaned seed in drills but broadcast berries or dried pulp. Seeds sown from berries have a slower and less successful

References
Elias, T.S. 1980. *The Complete Trees of North America Field Guide and Natural History.* New York: Van Nostrand Reinhold Company. 948p.

Foster, C.O. 1977. *Plants-a-Plenty: How to Multiply Outdoor and Indoor Plants Through Cuttings, Crown and Root Divisions, Grafting, Layering, and Seeds.* Emmaus, PA: Rodale Press, Inc. 328p.

Harris, A.S., and W.I. Stein. 1974. *Sorbus* L. Mountain-ash. pp. 780-84 *In*: Schopmeyer, C.S. (tech. coord.) 1974. *Seeds of the Woody Plants in the United States.* Agric. Handbook 450. Washington, DC: USDA Forest Service. 883p.

Kruckeberg, A.R. 1982. *Gardening with Native Plants of the Pacific Northwest.* Seattle, WA: University of Washington Press. 252p.

Matthews, R.F. 1993. *Sorbus sitchensis. In*: Fischer, William C. (comp.) The Fire Effects Information System [Monograph Online]. Missoula, MT: USDA Forest Service, Intermountain Fire Sciences Laboratory. http://www.fs.fed.us/database/feis/plants/Tree/SORSIT. Accessed June 25, 1996.

germination rate. When outplanting, two-year-old stock is preferred (Harris and Stein 1974, Matthews 1993).
Seeds per kilogram: ~145,500-385,805 (Harris and Stein 1974)
Vegetative: It has been reported that American mountain-ash (*S. americana*), a closely related species, sprouts from the stump when top-killed (Matthews 1993). Sitka mountain-ash is difficult to start from cuttings and is generally not successful (Foster 1977, Kruckeberg 1982).

Spiraea betulifolia
Birchleaf spirea

Description:

Birchleaf spirea, also called white spirea, is a deciduous, rhizomatous shrub that grows 15-60 cm high. The bark is scaly and cinnamon-brown in color. The leaves are alternate, 2.5-5 cm long, oval, and serrated on the ends. The white to pale pink flowers are found in a dense, flat-topped cluster and appear fuzzy due to many protruding stamens. These turn brown shortly after fertilization and yield the small, dry, podlike fruit that contain the seed. Birchleaf spirea is not an important browse species but is good for watershed rehabilitation (Habeck 1991, Strickler 1993).

Habitat and Geographic Range

Birchleaf spirea ranges from eastern Oregon and Washington through southern Idaho, east to north-central Wyoming, western Montana, and in the Black Hills of South Dakota. In Canada it occurs in southern British Columbia, southern Saskatchewan, and eastern Alberta. It grows from 300 to 1200 m in dry forests and up to 3000 m in wetter forests. It can be found on brushy or open slopes and scattered to dense forests. It grows on dry, rocky sites and in forest habitats with parent material ranging from limestone to quartz (Stickney 1974, Habeck 1991, Strickler 1993).

Propagation

Seed: Birchleaf spirea produces seed best when growing in full sun. The fruit ripens from mid-July to early September and seeds disseminate during October. The seed requires no stratification and can germinate at 0-2°C when conditions are maintained for more than 120 days. Seed can therefore be sown in the fall and overwintered (Stickney 1974).

References

Dirr, M.A., and C.W. Heuser. 1987. *The Reference Manual of Woody Plant Propagation: From Seed to Tissue Culture.* Athens, GA: Varsity Press, Inc. 239p.

Habeck, R.J. 1991. *Spiraea betulifolia. In*: Fischer, William C. (comp.) The Fire Effects Information System [Monograph Online]. Missoula, MT: USDA Forest Service, Intermountain Fire Sciences Laboratory. http://www.fs.fed.us/database/feis/plants/Shrub/SPIBET. Accessed June 27, 1996.

MSUE (Michigan State University Extension). Information Management Program (IMP) Information System. Home Horticulture. [Monograph Online]. http://lep.cl.msu.edu/msueimp/htdoc/impmsaic.html. Accessed June 27, 1996.

Stickney, P.F. 1974. *Spiraea betulifolia* Pall. var. *Lucida* (Dougl.) C.L. Hitchc. Birchleaf spirea. pp. 785-86 *In*: Schopmeyer, C.S. (tech. coord.) 1974. *Seeds of the Woody Plants in the United States*. Agric. Handbook 450. Washington, DC: USDA Forest Service. 883p.

Strickler, D. 1993. *Wayside Wildflowers of the Pacific Northwest*. Columbia Falls, MT: Flower Press. 272p.

Vegetative: Take cuttings in mid-June, dip in a 3000 ppm IBA talc, and stick into a sandy soil in an outdoor frame with bottom heat. Cuttings can also be taken in early August but survival is not as high (Dirr and Heuser 1987). Birchleaf spirea is rhizomatous and usually grows in extensive colonies. After a disturbance, birchleaf spirea will usually resprout within the next growing season. Buds are well distributed along the entire length of the rhizome. As a result, any section of the rhizome is probably capable of generating new stems if it is large enough to provide the carbohydrates necessary for sprouting. Birchleaf spirea also appears capable of layering from aerial stems (Habeck 1991).

Disease: Spireas can be infected by aphids, usually found on the shoot tips or in the flower clusters, which cause leaf curling and can be dislodged with a high-pressure water spray from a garden hose. Rolled and webbed leaves are a sign of oblique-banded leaf rollers which can be hand picked off infested leaves. Scales can also be a problem. Sprays of dormant oil in the spring will help minimize injury to predators that help control scales. Fire blight can infect spireas and the use of high nitrogen fertilizers makes them more susceptible. Symptoms include scorched-looking leaves, dieback of twig tips, and dead leaves hanging on blighted branches. There is no satisfactory chemical control for fire blight and infected branches should be pruned out (MSUE 1996).

Spiraea densiflora
Subalpine spirea

Description

Subalpine spirea is a small, deciduous, many-branched shrub that grows up to 1.2 m in height. The bark is reddish-brown in color. The leaves are alternate, oval to elliptical in shape, 2-4 cm long with coarsely toothed margins two-thirds of the way around. The tiny flowers are rose colored and borne in dense flat-topped to rounded clusters at the end of the stems. The stamens extend beyond the petals making the cluster appear fuzzy. The few-seeded fruit is a follicle found four to five per flower. The seeds are a food source for many wildlife species (Stead 1989, Taylor and Douglas 1995).

Habitat and Geographic Range

Subalpine spirea ranges from British Columbia south through the Olympic and Cascade Mountains, the Sierra Nevadas in California, and east to Idaho and Montana. It grows from 600 to 3050 m in elevation and can be found along streams and lakes, in forest clearings, on open rocky slopes, and in open subalpine forests. It grows in full sun to light shade in moist soils (Stead 1989, Taylor and Douglas 1995).

Propagation

Seed: Collect seed pods when they turn brown, allow them to air dry, and shake out the seed. Sow directly outdoors in the fall or cold stratify for spring planting. Soak dried seed in water for one or two days, place in a plastic bag with moist sand or peat, and store in a refrigerator for one to two months. Plant into 20-25-cm-deep containers in a sand, loamy soil, and peat moss mixture. Cover with a thin layer of soil and a light covering of mulch or plastic to prevent the seed from drying (Stead 1989).
Vegetative: Take softwood cuttings 7.6-15.2 cm long from the stem tips. Remove the lower leaves and dip the end into a rooting hormone. Stick cuttings into a container or propagating bed in a perlite:peat mixture and water well. Shade the cuttings and keep them under high humidity. They should root within four weeks and can be transplanted immediately (Stead 1989).

References

Stead, S. 1989. Mountain Spiraea. Plants for the Lake Tahoe Basin. Soil Conservation Service, Nevada Cooperative Extension Service. Fact Sheet 89-63.

Taylor, R.J,. and G.W. Douglas. 1995. *Mountain Plants of the Pacific Northwest: A Field Guide to Washington, Western British Columbia, and Southeastern Alaska.* Missoula, MT: Mountain Press Publishing Company. 437p

Spiraea douglasii
Douglas spirea

References

Darris, D.C., T.R. Flessner, and J.D.C. Trindle. 1994. Corvallis Plant Materials Center Technical Report: Plant Materials for Streambank Stabilization, 1980-1992. Portland, OR: USDA Natural Resources Conservation Service. 172p.

Esser, L.L. 1995. *Spiraea douglasii. In:* Fischer, William C. (comp.) The Fire Effects Information System [Monograph Online]. Missoula, MT: USDA Forest Service, Intermountain Fire Sciences Laboratory. http://www.fs.fed.us/database/feis/plants/Shrub/SPIDOU. Accessed June 27, 1996.

Helliwell, R. 1987. Forest Plants of the Warm Springs Indian Reservation. Warm Springs, OR: Confederated Tribes of the Warm Springs. 177p.

Description

Douglas spirea, also called hardhack, is a many-branched, erect shrub reaching 0.5-2 m in height. The leaves are alternate, oblong-elliptical, oblong-ovate, or elliptical in shape, 3-7.5 cm long, with serrated margins along the upper 0.6-1.3 cm. The small flowers are pink or reddish in color, borne in dense elongated terminal clusters, and appear fuzzy due to the many protruding stamens. The fruit is a smooth follicle that persists through the winter. The seeds are approximately 2 mm long (Esser 1995). Douglas spirea is browse for deer and occasionally for livestock. It is a good species for riparian revegetation programs (Helliwell 1987, Randall et al. 1994, Esser 1995).

Habitat and Geographic Range

Douglas spirea ranges from Alaska south into northern California and east to western Montana at elevations of 650-2050 m. It grows best on moist to semi-wet loam and sandy loam soils with good drainage. It can also be found growing on silty clay, clay loam, and gravelly substrates and is tolerant of permanently water-logged soils. It occurs on moist sites and riparian areas including wet meadows, floodplains, bogs, swamps, and along streams, rivers, lakes, and springs. It is generally a shade-intolerant species but frequently forms dense thickets in riparian areas (Esser 1995).

Propagation

Seed: Douglas spirea flowers from June to September. Fresh seed germinates easily without any pretreatment while dry seed may require one to two months of cold stratification (Darris et al. 1994).

Vegetative: Collect cuttings in the fall, making sure there is a bud on the length of cutting. Make a slice cut at the distal end. Place cuttings in a bucket of water in a cooler for a few weeks, drip them off, then place in a plastic bag in a cooler until April. Dip in a rooting hormone and plant in a peat:perlite (1:1) medium (Jebb 1995). Take hardwood cuttings, 15-20 cm long, during December and January, treat with a rooting hormone, and stick into a moist potting soil. Place cuttings in a greenhouse with an overhead mist system and bottom heat. After hardening off, outplant by late March or April (Darris et al. 1994). Douglas spirea is a rhizomatous shrub that often forms dense colonies. It will sprout from the stem base and root crown following disturbance (Esser 1995).

Disease: Aphids can infect spireas and are usually found on the shoot tips or in the flower clusters. They cause leaf curling and can be dislodged with a high-pressure water spray from a garden hose. Rolled and webbed leaves are a sign of oblique-banded leaf rollers, which can be hand picked off infested leaves. Scales can also be a problem. Sprays of dormant oil in the spring will help minimize injury to predators that help control scales. Fire blight can infect spireas and the use of high-nitrogen fertilizers makes them more susceptible. Symptoms include scorched-looking leaves, die back of twig tips, and dead leaves hanging on blighted branches. There is no satisfactory chemical control for fire blight and infected branches should be pruned out (MSUE 1996).

Jebb, T. 1995. Horticulturalist, USDI Bureau of Land Management, C.A. Sprague Seed Orchard, Merlin, OR. Personal communication.

MSUE (Michigan State University Extension). Information Management Program (IMP) Information System. Home Horticulture. [Monograph Online]. http://lep.cl.msu.edu/ msueimp/htdoc/ impmsaic.html. Accessed June 27, 1996.

Randall, W.R., R.F. Keniston, D.N. Bever, and E.C. Jensen. 1994. *Manual of Oregon Trees and Shrubs.* Corvallis, OR: Oregon State University Bookstores. 305p.

Symphoricarpos albus
Common snowberry

Description
Common snowberry is an erect, deciduous shrub, 0.5-2 m tall. The branches and leaves are opposite, with oval to elliptical shaped leaves. Flowers are tiny pinkish-white bells occuring in pairs. The fruit is a white drupe that grows in compact clusters and remains on the plant overwinter (Haeussler et al. 1990). The berries are an important food for pheasant, grouse, partridge, and quail. It is a good species for rehabilitating riparian areas and mine spoils (Snyder 1991).

Habitat and Geographic Range
Snowberry grows in upland, moist, well-drained soils, and can be found from sea level to mid-elevations in forests and on open slopes, and river banks. It ranges from Alaska east across Canada to Newfoundland, south to California, Nevada, and Utah, and east through the northern Midwest into New England. It is an excellent soil binder (Snyder 1991, King County 1994).

Propagation
Seed: Collect the seed from mid-October through winter by stripping or flailing the clusters of fruit onto canvas, or picking by hand. Clean by macerating with water and allowing the pulp to float away. Seed dried and stored in sealed containers at 5°C remains viable for up to two years. Warm stratify for 60 days at room temperature followed by 180 days of cold stratification at 5°C. Plant 0.6 cm deep in soil and mulch with a layer of sawdust (Evans 1974).
Seeds per kilogram: ~119,045-249,120 (Evans 1974)

References

Buis, S. 1996. Owner, Sound Native Plants, Olympia, WA. Personal communication.

Dirr, M.A., and C.W. Heuser Jr. 1987. *The Reference Manual of Woody Plant Propagation: From Seed to Tissue Culture*. Athens, GA: Varsity Press. 239p.

Evans, K.E. 1974. *Symphoricarpos* Duham. Snowberry. pp. 787-90 *In*: Schopmeyer, C.S. (tech. coord.) 1974. *Seeds of the Woody Plants in the United States*. Agric. Handbook 450. Washington, DC: USDA Forest Service. 883p.

Haeussler, S., D. Coates, and J. Mather. 1990. Autecology of common plants in British Columbia: A literature review. British Columbia Ministry of Forests. 272p.

Vegetative: Common snowberry propagates by woody runners just beneath the ground. Follow these to the parent plant, cut, and transplant into flats of perlite from October to February (King County 1994). Take hardwood cuttings, 15-20 cm long, from June to August and insert into ordinary soil in shade (Hellyer 1972). Store cuttings over winter in damp sawdust or peat moss. Dip in an IBA talc or solutions of 1000 to 3000 ppm (Dirr and Heuser 1987) and stick in potting soil in late February to early March. Common snowberry is very susceptible to powdery mildew; full sun and good air circulation help to control the problem (Buis 1996).

Hellyer, A.G.L. 1972. *Sanders' Encyclopaedia of Gardening*. London: Collingridge Books. 534p.

King County Department of Public Works, Surface Water Management Division. 1994. Northwest Native Plants, Identification and Propagation for Revegetation and Restoration Projects. King County, WA.

Snyder, S.A. 1991. *Symphoricarpos albus. In:* Fischer, William C. (comp.) The Fire Effects Information System [Monograph Online]. Missoula, MT: USDA Forest Service, Intermountain Fire Sciences Laboratory. http://www.fs.fed.us/database/feis/plants/Shrub/SYMALB. Accessed April 4, 1997.

Symphoricarpos oreophilus
Mountain snowberry

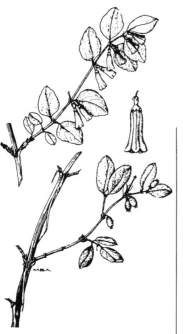

Description
Mountain snowberry is a deciduous shrub with low-growing, erect, and often trailing branches. It averages 0.6-1.2 m in height, but can reach up to 1.8 m on good sites. The stems are woody, many branched, often with grayish or reddish shreddy bark. The leaves are opposite, oval to elliptic in shape, slightly pubescent above, hairless to grayish pubescent below, with pointed tips and entire to dentate margins. The white or pink flowers are found in small clusters in the axils of the leaves or in terminal clusters. The fruit is a white egg-shaped berry 6-9 mm in diameter, and contains two seeds. The berry can persist on the plant for up to two seasons. Mountain snowberry has a fibrous root system and is good for establishing cover on bare and severe sites. It is an early spring browse for livestock and game and is used as cover by a variety of birds and small mammals (Van Dersal 1938, McMurray 1986, USDA 1988).

Habitat and Geographic Range
Mountain snowberry ranges from British Columbia to Montana, south to California, Texas, and northern Mexico. It can be found from the foothills into the subalpine zone. It is fairly shade intolerant and grows under open canopies, along the edges of meadows, and other openings. It grows in sandy to clay loam soils and occasionally on alluvial bottomlands and prefers moist well-drained soils. It is considered a climax species but establishes itself in the early seral stages (Van Dersal 1938, McMurray 1986).

Propagation
Seed: Seed matures from July through September depending on range and elevation. Collect by stripping or knocking into a hopper or container and clean by macerating in water, drying, and fanning. Seed should be stored dry and will remain viable for up to three years. Seed has a double dormancy and may need to soak in hot water prior to planting. Sow in the spring or sow unstratified seed in the fall. Sow at a moderate depth in well-drained soil. Young plants are very

References
McMurray, N.E. 1986. *Symphoricarpos oreophilus*. *In*: Fischer, William C. (comp.) The Fire Effects Information System [Monograph Online]. Missoula, MT: USDA Forest Service, Intermountain Fire Sciences Laboratory. http://www.fs.fed.us/database/feis/plants/Shrub/SYMORE. Accessed February 12, 1997.

Plummer, A.P., D.R. Christensen, and S.B. Monsen. 1968. Restoring Big Game Range in Utah. Utah Div. Fish Game Publ. 68-3. 183p.

USDA Forest Service. 1988. *Range Plant Handbook*. New York: Dover Publications, Inc. 816p.

sensitive to drought during their first year (Plummer et al. 1968, Vories 1980, McMurray 1986).

Seeds per kilogram: ~119,190 (Plummer et al. 1968)

Vegetative: Mountain snowberry has many means of vegetative reproduction. It is reported to be rhizomatous though the extent is unclear. It produces basal sprouts from a "root crown." However, it is unclear whether the structure is root or stem tissue. Perennating buds can be found 2-3 cm below the ground. Mountain snowberry has also been observed to reproduce by layering and stolon production (McMurray 1986).

Van Dersal, W.R. 1938. Native Woody Plants of the United States: Their Erosion-Control and Wildlife Values. USDA Misc. Pub. 303. 362p.

Vories, K.C. 1980. Growing Colorado Plants from Seed: A State of the Art. Vol. 1: Shrubs. Ogden, UT: USDA Forest Service Intermountain Forest and Range Experiment Station. Gen. Tech. Rep. INT-103. 80p.

Vaccinium deliciosum
Blueleaf huckleberry

References

Crowley, D.J. 1933. Observations and Experiments with Blueberries in Western Washington. State College of Washington. Agricultural Experiment Station. Bull. 276. 20p.

Gilkey, H.M., and L.J. Dennis. 1980. *Handbook of Northwestern Plants.* Corvallis, OR: Oregon State University Bookstores, Inc. 507p.

Hitchcock, C.L., and A. Cronquist. 1973. *Flora of the Pacific Northwest; An Illustrated Manual.* Seattle, WA: University of Washington Press. 730p.

Kruckeburg, A.R. 1982. *Gardening with Native Plants of the Pacific Northwest.* Seattle, WA: University of Washington Press. 252p.

Minore, D., and A.W. Smart. 1978. Frost tolerance in seedlings of *Vaccinium membranaceum, Vaccinium globulare,* and *Vaccinium deliciosum. Northwest Science* 52(3):179-85.

Peck, M.E. 1961. *A Manual of the Higher Plants of Oregon.* Portland, OR: Binfords & Mort Publishers. 936p.

Description

Blueleaf huckleberry, also known as Cascade huckleberry, is a low, bushy shrub growing 5-30 cm high. The branches are slightly angled; young branches are grayish and minutely pubescent, turning purplish with age. The leaves are obovate to oblanceolate, 2-4 cm long, finely dentate, very glaucous, and rounded to obtuse at the apex. The flowers are borne solitary on 3-6 mm long pedicles in the axils. The corolla is nearly globose and white to pinkish in color. The fruit is a glaucous-blue berry, 6-10 mm in diameter. Berries are eaten by humans, bears, and birds (Peck 1961, Gilkey and Dennis 1980, Taylor and Douglas 1995).

Habitat and Geographic Range

Blueleaf huckleberry ranges from southern British Columbia to northern Oregon and can be found in both the Cascade Mountain and the Olympic Mountain ranges. It grows on dry slopes at elevations above 1300 m (Peck 1961, Hitchcock and Cronquist 1973, Minore and Smart 1978).

Propagation

Seed: Collect berries in late summer or early fall. Clean by macerating in water, floating off the pulp, then allowing to dry. Seed requires no stratification and can be sown on a moist peat surface. Temperatures of 18°C (for 12 hours) during the day and 13°C (for 12 hours) at night are ideal for germination. Seven weeks after germination, change temperatures to 20°C (for 14 hours) and 14°C (for 10 hours). Fertilize seedlings 10 weeks after germination. After the seedlings are 12 weeks old, transplant to a peat:sand (1:1) media in individual pots (Minore and Smart 1978).

Vegetative: Blueleaf huckleberry can be propagated by cuttings, rooted suckers, and offshoots (Kruckeburg 1982). Take hardwood cuttings in January or February from two-year-old wood or older in lengths of about 15 cm. Plant horizontally 2.5 cm deep in a peat:sand (2:1) medium. New shoots appear around May or June and should be protected from direct sunlight. New roots should be established by the end of August and plants can be transplanted during the winter or spring. Take softwood cuttings as soon as the new growth becomes woody, usually around mid-June to mid-July. Use heel cuttings no longer than 10 cm and stick in a sand:peat medium out of direct sunlight. Most cuttings should be rooted by the end of October and can be transplanted to a nursery bed. Outplant to a permanent location one year after transplanting (Crowley 1933).

Taylor, R.J,. and G.W. Douglas. 1995. *Mountain Plants of the Pacific Northwest: A Field Guide to Washington, Western British Columbia, and Southeastern Alaska.* Missoula, MT: Mountain Press Publishing Company. 437p.

Vaccinium membranaceum
Black huckleberry

References

Albright, M. 1996. Greenhouse Manager. USDI National Park Service, Olympic National Park, Port Angeles, WA. Personal communication.

Dirr, M.A., and C.W. Heuser, Jr. 1987. *The Reference Manual of Woody Plant Propagation: From Seed to Tissue Culture.* Athens, GA: Varsity Press. 239p.

Haeussler, S., D. Coates, and J. Mather. 1990. Autecology of common plants in British Columbia: A literature review. B.C. Ministry of Forests. 272p.

Link, E. 1993. (ed.) Native Plant Propagation Techniques for National Parks: Interim Guide. East Lansing, MI: Rose Lake Plant Materials Center. 240p.

Minore, D., and A.W. Smart. 1978. Frost tolerance in seedlings of *Vaccinium membranaceum, Vaccinium globulare,* and *Vaccinium deliciosum. Northwest Science* 52(3):179-85.

Peck, M.E. 1961. *A Manual of the Higher Plants of Oregon.* Portland, OR: Binfords & Mort. 936p.

Description

Black huckleberry is an erect, deciduous shrub 0.1-2 m tall. The leaves are 2.5-5 cm long, elliptical to oblong with a long pointed tip and a finely serrated margin. The bell-shaped flowers are pinkish and found singly on the underside of the twigs. The fruit is a round, purplish-black berry 6-8 mm in diameter. The fruit is eaten by many birds, bears, and small mammals, while deer, elk, rabbit and other small mammals browse the twigs and foliage (Haeussler et al. 1990).

Habitat and Geographic Range

Black huckleberry is found on dry to moist sandy or gravelly loams (Haeussler 1990). It grows at 1000-1800 m from British Columbia south to California, and east to Michigan (Peck 1961, Minore and Smart 1978).

Propagation

Seed: Collect the berries in the fall and clean by running them through a blender with dull blades, straining the pulp with a sieve, and spreading them to dry on a paper towel (Minore and Smart 1978, Link 1993). According to Haeussler et al. (1990), seeds require no stratification or scarification and germinate within 16-21 days of sowing. However, Albright (1996) has found poor germination without stratification and recommends overwintering of seeds in flats outside. Germination can also be achieved by sowing the seed on moist peat in a growth chamber at 18°C (for 12 hours a day) and 13°C (for 12 hours a day). Seven weeks after germination, warm the growth chamber to 20°C (for 14 hours a day) and 14°C (for 10 hours). Fertilize the seedlings after they are 10 weeks old and transplant into a peat:sand mixture (1:1) in individual pots after 12 weeks (Minore and Smart 1978). **Vegetative:** Take cuttings from rhizomes in early spring or late summer and autumn by digging up the rhizomes, cutting them into lengths of 10 cm, and placing them in vermiculite at 21°C (Dirr and Heuser 1987, Haeussler et al. 1990).

Vaccinium parvifolium
Red huckleberry

Description

Red huckleberry is a deciduous shrub that grows from 1 to 3 m in height with dense, fine branches. The leaves are alternate, simple, oval to elliptical, grayish to green, glaucuous above, with entire margins. The urn-shaped flowers are small and greenish white in color and occur singly in the leaf axils. The fruit is a translucent, red, fleshy berry, 8 mm in diameter which contains eighteen to nineteen seeds (Tirmenstein 1990). Red huckleberry is an important food source for wildlife. The berries are eaten by birds and mammals and the shoots and foliage are browsed by elk and deer (Randall et al. 1994).

Habitat and Geographic Range

Red huckleberry grows on a variety of sites from Alaska south to California (Randall et al. 1994). It prefers light shade in coniferous forests and dry to slightly moist soils (King County 1994). It ranges from sea level to 1500 m (Randall et al. 1994).

Propagation

Seed: Fruit ripens from July to August and is easily collected by hand picking or by beating the bush over a large bucket. Following collection, chill the fruit at 10°C for several days. Clean by macerating and floating off the pulp and unsound seeds. Seeds dried at 15-21°C for two days can be stored in a refrigerator for up to 12 years. Stored seed germinates well when exposed to alternating temperature and light regimes of 28°C light for 14 hours a day and 13°C dark for 10 hours (Tirmenstein 1990). Plant in a mixture of sand and peat moss, and transplant after seven weeks, then outplant after the first growing season (Crossley 1974).
Seeds per kilogram: ~5,268,955-7,142,860 (Crossley 1974)
Vegetative: Red huckleberry sprouts after plants are damaged. Take cuttings during the dormant season (Crossley 1974).

References

Crossley, J.A. 1974. *Vaccinium* L. Blueberry. pp. 840-43 *In*: Schopmeyer, C.S. (tech. coord.) 1974. *Seeds of the Woody Plants in the United States*. Agric. Handbook 450. Washington, DC: USDA Forest Service. 883p.

King County Department of Public Works, Surface Water Management Division. 1994. Northwest Native Plants, Identification and Propagation for Revegetation and Restoration Projects. King County, WA.

Randall, W.R., R.F. Keniston, D.N. Bever, and E.C. Jensen. 1994. *Manual of Oregon Trees and Shrubs*. Corvallis, OR: Oregon State University Bookstores. 305p.

Tirmenstein, D.A. 1990. *Vaccinium parvifolium. In*: Fischer, William C. (comp.) The Fire Effects Information System [Monograph Online]. Missoula, MT: USDA Forest Service, Intermountain Fire Sciences Laboratory. http://www.fs.fed.us/database/feis/plants/Shrub/VACPAR. Accessed April 4, 1997.

References

Crossley, J.A. 1974. *Vaccinium* L. Blueberry. pp. 840-43 *In*: Schopmeyer, C.S. (tech. coord.) 1974. *Seeds of the Woody Plants in the United States*. Agric. Handbook 450. Washington, DC: USDA Forest Service. 883p.

Dirr, M.A., and C.W. Heuser. 1987. *The Reference Manual of Woody Plant Propagation: From Seed to Tissue Culture*. Athens, GA: Varsity Press, Inc. 239p.

Peck, M.E. 1961. *A Manual of the Higher Plants of Oregon*. Portland, OR: Binfords & Mort. 936p.

Romme, W.H., L. Bohland, C. Persichetty, and T. Caruso. 1995. Germination ecology of some common forest herbs in Yellowstone National Park, Wyoming, U.S.A. *Arctic and Alpine Research* 27(4):407-12.

Tirmenstein, D. 1990. *Vaccinium scoparium*. *In*: Fischer, William C. (comp.) The Fire Effects Information System [Monograph Online]. Missoula, MT: USDA Forest Service, Intermountain Fire Sciences Laboratory. http://www.fs.fed.us/database/feis/plants/Shrub/ARCVIS. Accessed March 12, 1997.

USDA Forest Service. 1988. *Range Plant Handbook*. New York: Dover Publications, Inc. 816p.

Vaccinium scoparium
Grouse huckleberry

Description

Grouse huckleberry is a dwarfed shrub reaching 10-50 cm in height. The elongated branches are slender, bright green when young, graying with age, and sharply angled. The leaves are alternate, deciduous, thin, narrowly ovate, 8-16 mm long, pointed at both ends, with finely serrate margins. The flowers are pink or white, nodding, and found solitary in the axils. The fruit is a bright red globose berry 3-4 mm in diameter and contains ten to twenty yellowish-brown seeds. Grouse huckleberry is not a favorable browse. The berries are eaten by birds, mammals, and humans. However, berries are rarely gathered due to their size and the limited number produced (Peck 1961, USDA 1988, Tirmenstein 1990).

Habitat and Geographic Range

Grouse huckleberry ranges from Alaska south to California, northern New Mexico, the Rocky Mountains, and Alberta. It can be found from 760 to 2300 m in the Pacific Northwest and 2600-3800 m in Colorado. It grows in acidic soils in both moist and dry sites, but is most common in sandy or gravelly loams and is almost always found in the understory of lodgepole pine (*Pinus contorta*) stands (Peck 1961, USDA 1988, Tirmenstein 1990).

Propagation

Seed: The berries ripen from late July through September. After collection, place the berries in a plastic bag and keep them at 5°C from a few days to a few weeks. Clean seed by macerating and floating the pulp and unsound seed off the top. Seed should be dried before storing. One study reported best germination with a cold stratification with warm night temperatures of 13°C. It is also reported to need no pretreatment (Crossley 1974, Dirr and Heuser 1987, USDA 1988, Romme et al. 1995).

Vegetative: Grouse huckleberry can be propagated by rhizome cuttings (Tirmenstein 1990).

Viburnum edule
Highbush cranberry

Description

Highbush cranberry is a multi-stemmed deciduous shrub growing 0.5-3 m tall. It has several stems that may grow to 4 cm in diameter. The leaves are opposite, ovate to obovate in shape, with three shallow, serrated lobes. The white flowers are in a small, dense, rounded cluster. The fruit is an orange to red drupe that contains one seed and can remain on the plant over the winter (Haeussler et al. 1990). Highbush cranberry fruit is consumed by bears, small mammals, and songbirds. Foliage is browsed by beaver, rabbit, and snowshoe hare (Matthews 1992).

Habitat and Geographic Range

Highbush cranberry can be found in moist, well-drained sites under hardwoods or mixed hardwood and softwoods (Govt. of Saskatchewan 1989). It is distributed throughout Alaska and across Canada to Newfoundland, south through New Enlgand, the Great Lakes states, and the Pacific Northwest at elevations from sea level to 1500 m (Matthews 1992). It is semi-tolerant to shade tolerant (Haeussler et al. 1990).

Propagation

Seed: Highbush cranberry begins producing an annual seed crop when the plant is approximately five years old. Clean, dry seed is viable in storage for up to ten years. Seed is difficult to germinate due to both seed coat and embryo dormancy. Best germination occurs when a warm stratification is followed by a cold stratification (Gill and Pogge 1974).
Seeds per kilogram: ~20,720-39,240 (Gill and Pogge 1974)
Vegetative: Take cuttings from one-year-old and new growth during July and August. Treat with a rooting hormone powder and stick into flats of perlite. Use intermittent mist until rooting occurs (about eight weeks), then transplant into flats of potting soil (Jarussi and Holloway 1987). Highbush cranberry also reproduces vegetatively by layering and sprouting from damaged root stocks, stembases, and stumps (Matthews 1992).

References

Gill, J.D., and F.L. Pogge. 1974. *Viburnum* L. Viburnum. pp. 844-50 *In*: Schopmeyer, C.S. (tech. coord.) 1974. *Seeds of the Woody Plants in the United States*. Agric. Handbook 450. Washington, DC: USDA Forest Service. 883p.

Government of Saskatchewan. 1989. Guide to Forest Understory Vegetation in Saskatchewan. Canada Forestry. Tech. Bull. No. 9/1980 revised January, 1989. 106p.

Haeussler, S., D. Coates, and J. Mather. 1990. Autecology of common plants in British Columbia: A literature review. B.C. Ministry of Forests. 272p.

Jarussi, R., and P.S. Holloway. 1987. Propagation of highbush cranberry, *Viburnum edule*, by stem cuttings. pp. 92-94 *In*: Proceedings, sixth annual Alaska Greenhouse and Nursery Conference. Palmer, AK; Feb. 11-12, 1987.

Matthews, R.F. 1992. *Viburnum edule. In*: Fischer, William C. (comp.) The Fire Effects Information System [Monograph Online]. Missoula, MT: USDA Forest Service, Intermountain Fire Sciences Laboratory. http://www.fs.fed.us/database/feis/plants/Shrub/VACPAR. Accessed April 4, 1997.

Trees

Abies grandis
Grand fir

References

Crane, M.F. 1991. *Abies grandis*. *In*: Fischer, William C. (comp.) The Fire Effects Information System [Monograph Online]. Missoula, MT: USDA Forest Service, Intermountain Fire Sciences Laboratory. http://www.fs.fed.us/database/feis/plants/Tree/ABIGRA. Accessed July 10, 1997.

Franklin, J.F. 1974. *Abies* Mill. Fir. pp. 168-83 *In*: Schopmeyer, C.S. (tech. coord.) 1974. *Seeds of the Woody Plants in the United States*. Agric. Handbook 450. Washington, DC: USDA Forest Service. 883p.

Description

Grand fir is an evergreen conifer tree which reaches 40 to 60 m in height and 50-100 cm in diameter. The bark is grayish to light brown, smooth or shallowly ridged, and flakes on mature trees. The dark green needles are long, blunt-tipped, and whitish underneath and are arranged in two rows on opposite sides of the stem. Needles are 2-5 cm long, linear, flat, rounded and notched at the tip, grooved on the upper side, with two white bands on the underside, tend to curve upward at the top of the tree, and are spread out flat on the lower branches (Crane 1991, Pojar and MacKinnon 1994, Randall et al. 1994). The female cones stand erect on the upper branches of the tree while the pollen-bearing male cones hang from the lower branches. Although the wood is weaker than that of other conifer species, it is a valuable commercial timber species and is used for pulp and light-duty uses (Crane 1991). Grand fir is also grown for Christmas trees. It provides good cover for wildlife. Grouse feed on the needles and birds, squirrels, and other small mammals eat the seeds (Crane 1991).

Habitat and Geographic Range

Grand fir is found in western Montana and northern Idaho to southern British Columbia and south to eastern Oregon and the northern coast of California (Franklin 1974). It grows in stream bottoms and valleys, and on mountain slopes and can be found from sea level up to 1830 m (Randall et al. 1994). It grows best on deep, nutrient-rich soils but will also grow on sandstone, weathered lava, granite, or gneiss-derived soils (Crane 1991).

Propagation

Seed: Grand fir begins producing seed at 20-30 years of age. Flowering occurs from mid-April to mid-June, cones ripen in August, and seed dispersal is in late August to early September. Collect cones by hand after they are ripened but before seed dispersal. Take care if climbing trees to collect cones since *Abies* stems can be relatively brittle. Seed require a period of after-ripening in the cone and should not be extracted immediately

after collection. Store cones in mesh sacks for several weeks or months at moderate temperatures and low humidity. Separate seeds by breaking up the dried cones, then tumbling, shaking, and screening. Seeds can be dewinged by hand or mechanically but take care not to cause damage. Seed can be stored dry in sealed containers at -5°C for five or more years. Seed should undergo a cold, moist stratification for 15-30 days. Sow stratified seeds in spring at a depth of 0.5 cm and cover with a thin layer of mulch (Franklin 1974).
Seeds per kilogram: ~26,230-63,495 (Franklin 1974)

Pojar, J., and A. MacKinnon. 1994. *Plants of the Pacific Northwest Coast: Washington, Oregon, British Columbia, and Alaska*. Vancouver, BC, Canada: British Columbia Ministry of Forests and Lone Pine Publishing. 527p.

Randall, W.R., R.F. Keniston, D.N. Bever, and E.C. Jensen. 1994. *Manual of Oregon Trees and Shrubs*. Corvallis, OR: Oregon State University Bookstores. 305p.

Acer macrophyllum
Bigleaf maple

Description

Bigleaf maple is a long-lived, large deciduous tree reaching 30 m in height, with a broad crown. The bark becomes furrowed with small plates, and is gray to reddish-brown in color. The leaves are 15-30 cm in length and width, have five deep lobes, and are pointed at the tips and heart shaped at the base, with few blunt irregular teeth along the margin. The flowers are greenish-yellow and hang in elongated clusters at the end of the twigs. The fruit is a winged samara at an angle of 60° or less. The seeds are eaten by squirrels, chipmunks, mice, and many birds, while the saplings, young twigs, and leaves are eaten by blacktail deer, mule deer, and elk. Several species of perching birds use bigleaf maples as nesting sites. Bigleaf maple is an important source of hardwood lumber. The wood is hard, but not very strong, and is used for furniture, paneling, cabinets, veneer, and musical instruments. Although not produced commercially, maple syrup can be made from the sap. It is also a good tree to plant along streambanks to prevent erosion (Elias 1980, Uchytil 1989).

Habitat and Geographic Range

Bigleaf maple grows from sea level to 2000 m in elevation and can be found growing in a variety of soils from deep and loamy to shallow and rocky, although it attains its best development on deep alluvial soils near streams. It is extremely flood tolerant and often remains in floodplain habitats. Its range is from southwestern British Columbia to southern California. It is usually found in association with Douglas-fir, western hemlock, and vine maple, but can form pure stands. Bigleaf maple's shade tolerance is low to moderate. It grows most rapidly in small forest openings and open areas (Elias 1980, Uchytil 1989).

Propagation

Seed: Flowers are first produced at about ten years of age. Trees growing in open habitats produce seed at an earlier age and in larger quantities than trees growing in shade. Seed production is often heavy. Fruit ripens

References

Buis, S. 1996. Owner, Sound Native Plants, Olympia, WA. Personal communication.

Elias, T.S. 1980. *The Complete Trees of North America Field Guide and Natural History.* New York: Van Nostrand Reinhold Company. 948p.

Haeussler, S., D. Coates, and J. Mather. 1990. Autecology of common plants in British Columbia: A literature review. British Columbia Ministry of Forests. FRDA Report-158. 272p.

Olson, D.F. Jr., and W.J. Gabriel. 1974. *Acer* L. Maple. pp. 187-94 *In*: Schopmeyer, C.S. (tech. coord.) 1974. *Seeds of the Woody Plants in the United States.* Agric. Handbook 450. Washington, DC: USDA Forest Service. 883p.

during September and October and is dispersed from October through January. Seed tends to decay rapidly and cannot be stored for long periods of time. Zasada et al. (1990) suggest collecting the seeds as late as possible in the fall but before rains begin; moisture content should be at a minimum at this time. Place the seeds in airtight containers soon after collection and store at 1°C until stratification begins. Cold stratify at 1-5°C for 40-80 days prior to sowing. Buis (1996) suggests stratifying over winter in a refrigerator and sowing in February or early March but has also noted excellent germination by sowing directly in the fall. Sow in mulched beds and grow for two years before transplanting or outplanting (Olson and Gabriel 1974, Uchytil 1989, Haeussler at al. 1990).

Seeds per kilogram: ~5,950-8,820 (Olson and Gabriel 1974)

Vegetative: Bigleaf maple sprouts vigorously from the root crown after it is top killed or cut (Uchytil 1989). Small seedlings, with about two or three leaves, can be salvaged from construction sites or from under mature trees and transplanted into containers (Buis 1996).

Uchytil, R.J. 1989. *Acer macrophyllum. In:* Fischer, William C. (comp.) The Fire Effects Information System [Monograph Online]. Missoula, MT: USDA Forest Service, Intermountain Fire Sciences Laboratory. http://www.fs.fed.us/database/feis/plants/Tree/ACEMAC. Accessed June 25, 1996.

Zasada, J.C., J.C. Tappeiner, II., and T.A. Max. 1990. Viability of bigleaf maple seeds after storage. *Western Journal of Applied Forestry* 5:52-55.

Alnus incana
Thinleaf alder (mountain alder)

References

Haeussler, S., D. Coates, and J. Mather. 1990. Autecology of common plants in British Columbia: A literature review. British Columbia Ministry of Forests. FRDA Report-158. 272p.

Java, B.J., and R.L. Everett. 1992. Rooting hardwood cuttings of Sitka and thinleaf alder. pp. 138-41 *In*: Proceedings, Symposium on Ecology and Management of Riparian Shrub Communities. USDA Forest Service. Gen. Tech. Rep. INT-289.

Schopmeyer, C.S. 1974. *Alnus* B. Ehrh. Alder. pp. 206-11 *In*: Schopmeyer, C.S. (tech. coord.) 1974. *Seeds of the Woody Plants in the United States*. Agric. Handbook 450. Washington, DC: USDA Forest Service. 883p.

Description

Thinleaf alder is a deciduous, shade-tolerant, multistemmed shrub or small tree which tends to form thickets and typically grows 2-5 m tall, but can reach heights of up to 15 m. The bark is thin, smooth, and grayish. The leaves are broadly elliptic or ovate-oblong, 3-7 cm long, dull green, with doubly dentate margins. The male catkins are 3-10 cm long and are clustered at the end of the twig. The female catkins are 9-13 mm long, semiwoody, conelike, and are on short, stout stalks in clusters of three to nine. The single-seeded nutlets have narrow lateral wings and are borne in conelike fruits which remain on the plant for about a year after seeds are shed. Thinleaf alder is a nitrogen fixer and has a high flood tolerance and thus improves soil fertility and stabilizes streambanks. The twigs and leaves of younger seedlings are eaten by deer, elk, moose, muskrats, cottontails, and snowshoe hares. Beavers eat the bark and build dams and lodges with the stems. Alder seeds, buds, and catkins are eaten by many birds and are considered an important winter food source (Uchytil 1989, Haeussler et al. 1990).

Habitat and Geographic Range

Thinleaf alder is the most widely distributed alder in western North America and is the most common alder of the Rocky Mountains, the Sierra Nevada, and the east side of the Cascades. It is found on a wide variety of sites, from near sea level to 3000 m. Thinleaf alder ranges from central Alaska and the Yukon Territory, southeast to western Saskatchewan and British Columbia, and south to New Mexico and California. It seldom grows away from water and is typically found near streams, rivers, or springs on moist mountain slopes on poorly developed soils of cobbles, gravels, or sands which remain moist year round due to high water tables (Uchytil 1989).

Propagation

Seed: Thinleaf alder produces abundant seed which is dispersed during fall and winter. However, seed viability can be quite low (Uchytil 1989). A seed crop is produced every one to four years. The fruit ripens from September to November and should be collected when the bracts start to separate. They will fully open on a drying rack in a well-ventilated room if left for several weeks at ambient air temperature. Fresh seeds will germinate without stratification. Seed can remain viable in storage for up to ten years when stored in sealed containers at 1-3°C. Stratify dried seed for 180 days at 5°C. Spring sowing works best and outplanting is usually done with one-year-old stock (Schopmeyer 1974).

Seeds per kilogram: ~963,400-1,984,130 (Schopmeyer 1974)

Vegetative: Exposed roots in streams will sprout and submerged branches sometimes form adventitious roots. Like other alders, thinleaf alder can sprout from the root collar or stump if damaged (Uchytil 1989, Haeussler et al. 1990). The rooting ability of cuttings can be enhanced by treating the basal end with solutions of 2000 ppm of IBA and 1000 ppm of naphthalene acetic acid then placing them in cold storage at 1-3°C for one month (Java and Everett 1992).

Uchytil, R.J. 1989. *Alnus incana* ssp. *tenuifolia*. In: Fischer, William C. (comp.) The Fire Effects Information System [Monograph Online]. Missoula, MT: USDA Forest Service, Intermountain Fire Sciences Laboratory. http://www.fs.fed.us/ database/feis/plants/ Tree/ALNINC. Accessed June 27, 1996.

Alnus rubra
Red alder

References

Elias, T.S. 1980. *The Complete Trees of North America Field Guide and Natural History.* New York: Van Nostrand Reinhold Company. 948p.

Harlow, W.M., E.S. Harrar, and F.M. White. 1979. *Textbook of Dendrology Covering the Important Forest Trees of the United States and Canada.* Sixth edition. New York: McGraw-Hill Book Company. 510p.

Hibbs, D.E., and A.A. Ager. 1989. Red Alder: Guidelines for Seed Collection, Handling, and Storage. Corvallis, OR: Forest Research Laboratory, Oregon State University. Special Publication 18. 6p.

Description

Red alder is a deciduous tree growing up to 35 m in height. The leaves are alternate, 7.5-13 cm long, ovate shaped, with a double-serrated margin. The bark is smooth and bluish-gray in color. The male and female catkins occur separately on the same tree. Flowers are produced in late summer and are borne in clusters of catkins. The male catkins are thin and 10-15 cm long while the female catkins are thicker and shorter, 0.8-1.2 cm long. The seeds are tiny winged nutlets found inside woody conelike fruits called strobiles. The strobiles are 1.3-2.5 cm long, cylindrical shaped, persistent, and turn from green or yellow to brown as they ripen. Red alder is one of the first species to appear after wildfires, logging, and other disturbances. It is a good nitrogen fixer. It is not a significant wildlife tree although its leaves and twigs are browsed by deer and elk. Red alder is an important hardwood in the Pacific Northwest, used mostly for furniture and veneers (Harlow et al. 1979, Elias 1980, Jensen et al. 1995).

Habitat and Geographic Range

Red alder grows in moist, rich, loamy soils, particularly in bottomlands and along streams. It is a lowland tree rarely growing above 1000 m in elevation and is usually found within 80 km of salt water. Its range is from southern Alaska to northern California. It can be found in pure stands but is more commonly found with Douglas-fir, Sitka spruce, western red cedar, western hemlock, Oregon ash, black cottonwood, bigleaf and vine maples, and Pacific dogwood (Elias 1980).

Propagation

Seed: Red alder seed ripens between early September and mid-October, depending on geographic location. Seed maturity can be checked by twisting a strobile along the long axis. If it is mature, the strobile will twist easily and the scales part slightly. Collect strobiles by stripping from the branches with a cone rake and blow out any leaf litter with a fan. Strobiles found on the upper third of the tree usually contain the most viable seed. Once picked, dry the strobiles immediately with good air circulation (on screens or in mesh bags at 16-27°C) to prevent molding. Extract seeds by tumbling and clean by passing them through a screen. Remove small material by processing the seed with an air column. Once cleaned, seed can be stored for a short time by refrigeration. For long-term storage, dry to less than 10% moisture content and store in moisture-proof containers at -12 to -13°C for up to five years. No stratification is required for red alder seed. The best time to sow seeds is in spring. Due to their size, seeds should not be covered or they will not germinate. Red alder seeds require plenty of water and a low-nitrogen, high-phosphorous fertilizer for germination and growth (Hibbs and Ager 1989, Jebb 1995).

Seeds per kilogram: ~844,355-2,396,385 (Schopmeyer 1974)

Vegetative: Red alder softwood cuttings are best if taken from shoots of trees less than seven years in age or from epicormic sprouts of older trees. Cuttings should be 6-12 cm long and 2-4 mm in diameter. A chemical treatment of 2000-4000 ppm IBA solution increases rooting. Plant cuttings 2-4 cm deep in a moist perlite and vermiculite (1:1) media and maintain at 25°C until rooted. Once rooted, transplant the cuttings to a vermiculite, perlite, and sandy loam (1:1:1) mixture and keep in a greenhouse for approximately five weeks with day and night temperatures of 21°C and 15°C, respectively. Then move them to a growth chamber and allow them to harden off before storing in a cold room during the winter and outplanting in the spring (Radwan et al. 1989). In addition, small seedlings can be collected from beneath mature trees and potted into containers.

Jebb, T. 1995. Horticulturalist, USDI Bureau of Land Management. C.A. Sprague Seed Orchard, Merlin, OR. Personal communication.

Jensen, E.C., D.J. Anderson, and D.E. Hibbs. 1995. The Reproductive Ecology of Broadleaved Trees and Shrubs: Red Alder, *Alnus rubra* Bong. Corvallis, OR: Forest Research Laboratory, Oregon State University. Research Publication 9c. 7p.

Radwan, M.A., T.A. Max, and D.W. Johnson. 1989. Softwood cuttings for propagation of red alder. *New Forests* 3:21-30.

Schopmeyer, C.S. 1974. *Alnus* B. Ehrh. Alder. pp. 206-11 *In*: Schopmeyer, C.S. (tech. coord.) 1974. *Seeds of the Woody Plants in the United States*. Agric. Handbook 450. Washington, DC: USDA Forest Service. 883p.

Alnus sinuata
(A. viridis spp. sinuata)
Sitka alder

References

Darris, D.C., T.R. Flessner, and J.D.C. Trindle. Corvallis Plant Materials Center Technical Report: Plant Materials for Streambank Stabilization, 1980-1992. Portland, OR: USDA Natural Resources Conservation Service. 172p.

Elias, T.S. 1980. *The Complete Trees of North America Field Guide and Natural History.* New York: Van Nostrand Reinhold Company. 948p.

Uchytil, R.J. 1989. *Alnus viridis* ssp. *sinuata. In*: Fischer, William C. (comp.) The Fire Effects Information System [Monograph Online]. Missoula, MT: USDA Forest Service, Intermountain Fire Sciences Laboratory. http://www.fs.fed.us/database/feis/plants/Tree/ALNVIR. Accessed June 27, 1996.

Description

Sitka alder is a deciduous shrub or small tree that grows about 10 m in height. The trunk rarely reaches more than 25 cm in diameter, and often grows crooked. The leaves are alternate, 7-14 cm long and 3-10 cm wide. Margins are doubly serrated. Female catkins, 0.7-1 cm long, are found in long clusters on long stalks. Male catkins are 10-14 cm long and are found hanging at the end of the branch. The fruit is a small, single-seeded nutlet with wide lateral wings. Sitka alder is browsed by deer and the seeds are eaten by small birds. Sitka alder is too small for lumber, but the wood is used for smoking fish and firewood. Due to its nitrogen-fixing ability, it improves soil fertility. Sitka alder is one of the first trees present after disturbances such as logging, fire, and landslides. It also helps to prevent soil erosion (Elias 1980).

Habitat and Geographic Range

Sitka alder is a fast-growing tree that grows from sea level up to 1200 m in elevation. It grows in rocky or gravelly soils, and can be found along streams, rivers, and swamps, and in marshy flats. It ranges from southern Alaska south to northern California and into northern Idaho and Montana (Elias 1980).

Propagation

Seed: Sitka alder begins producing seed at four to seven years of age. Seeds are dispersed during the fall and require a cold, moist stratification at 3°C for one to three months. Dusting with a fungicide will help prevent mold but can reduce final germination percentage. Plants grow best in mineral soil with regular water and full sunlight (Uchytil 1989, Darris et al. 1994).
Vegetative: Sitka alder plants can sprout from the root collar or stump when damaged. Propagation by stem cuttings of Sitka alder is generally not very successful (Uchytil 1989).

Arbutus menziesii
Pacific madrone

Description

Pacific madrone is a broadleaved evergreen tree that grows 9-12 m in its northern range and 24-40 m in the redwood forests of northwest California. It has a broad, spreading crown with crooked, twisting branches. Older bark is reddish-brown and peels off while young bark is smooth and yellowish-green. The leaves are alternate, oval, leathery, 7.5-13 cm long, dark green and shiny above, whitish-green below, many veined, with entire margins except for occasional fine teeth on young leaves. Leaves are shed in May to late summer of their second year. The flowers are white, urn shaped, 6-12 mm long, borne in dense terminal clusters. The fruit is an orange-red round berry, 1-1.5 cm in diameter with a granular surface, and contains numerous dark-brown seeds. The wood is occasionally used in flooring and cabinet making. The fruit is eaten by many types of wildlife and the tree is used by both open-nesting and cavity-nesting birds (Elias 1980, McMurray 1989, Pojar and MacKinnon 1994).

Habitat and Geographic Range

Pacific madrone ranges from southwestern British Columbia south to the Coast Ranges of southern California. It is more common west of the Cascade Mountains in Oregon and Washington and is found scattered along the west slopes of the Sierra Nevadas in central California. Its elevational range is from sea level to 1800 m. In general, Pacific madrone inhabits areas with rocky, fine-textured soils with low moisture content in the summer. In the southern portion of its range, it can be found growing on dry foothills, wooded slopes, and sunny, rocky sites. In the north, it grows in hot, dry lowland sites and is generally found in areas with mild oceanic winters. It is a moderately shade-tolerant plant (Elias 1980, McMurray 1989).

Propagation

Seed: Pacific madrone is a prolific seed producer. The fruit ripens in September and October and remains on the tree until December. Collect berries from October through December. Lay them out to dry or macerate

References

Elias, T.S. 1980. *The Complete Trees of North America Field Guide and Natural History.* New York: Van Nostrand Reinhold Company. 948p.

McMurray, N.E. 1989. *Arbutus menziesii. In*: Fischer, William C. (comp.) The Fire Effects Information System [Monograph Online]. Missoula, MT: USDA Forest Service, Intermountain Fire Sciences Laboratory. http://www.fs.fed.us/database/feis/plants/Tree/ARBMEN. Accessed March 4, 1997.

Pojar, J., and A. MacKinnon. 1994. *Plants of the Pacific Northwest Coast: Washington, Oregon, British Columbia, and Alaska.* Vancouver, BC, Canada: British Columbia Ministry of Forests and Lone Pine Publishing. 527p.

Prockter, N.J. 1976. *Simple Propagation: Propagating by Seed, Division, Layering, Cuttings, Budding and Grafting.* London: Faber and Faber. 246 p.

Roy D.F. 1974. *Arbutus menziesii* Pursh. Pacific madrone. pp. 226-27 *In*: Schopmeyer, C.S. (tech. coord.) 1974. *Seeds of the Woody Plants in the United States.* Agric. Handbook 450. Washington, DC: USDA Forest Service. 883p.

them and float off the pulp. Dried seed or berries can be stored at room temperature for one or two years or in airtight containers at 1-4°C for longer periods of time. Seed should undergo a moist stratification at 0.5-4°C for 30-90 days or stratified naturally outdoors over winter. Use of sulfuric acid prior to stratification has not been found to increase germination. Germinate seed in flats in a sand-peat medium and transplant to individual containers when large enough to handle. Pacific madrone seedlings grow very slowly. Average height of second-year seedlings is approximately 2.5 cm (Roy 1974, Prockter 1976, McMurray 1989).

Seeds per kilogram: ~434,300-705,470 (Roy 1974)

Vegetative: The base of the Pacific madrone stem has a woody, underground regenerative burl (McMurray 1989) and sprouts readily following disturbance. This species can be propagated from both cuttings and layering (Roy 1974).

Betula papyrifera
Paper birch

Description

Paper birch is a medium-sized, deciduous tree that grows up to 25 m tall with a rounded crown. The distinctive bark is cream-white with small, dark, horizontal lenticels and peels readily in sheets. The leaves are alternate, triangular to ovate shaped, pointed at the tip, and serrated along the margin except close to the base of the leaf. The male and female flowers occur as separate catkins on the same tree and are found near the tips of the branchlets. The male catkins are 7-10 cm long while the female are 2.5-3 cm long. The fruits are a drooping, catkinlike strobile consisting of an abundance of scales each enclosing a tiny winged seed. The twigs are a food source for moose and deer, and beaver feed on the inner bark. The seeds are eaten by small birds and rodents. The wood is strong and hard and is used for veneer and pulpwood. The bark is used for birchbark canoes and the sap can be boiled down for syrup (Elias 1980).

Habitat and Geographic Range

Paper birch grows at moderate elevations in a wide range of soil and moisture conditions but prefers a well-drained, sandy loam soil. It may form pure stands on wet sites. It ranges across Canada southward into the northern United States. It is a pioneer tree after fire disturbance and is commonly associated with white and black spruce and poplar (Elias 1980).

Propagation

Seed: Paper birch is a prolific seed producer, beginning at about age fifteen, with optimum production at forty to seventy years of age. Good seed crops are produced every second to third year. Nearly all the seed (about 90-95%) is shed from September through November. Between 14 and 47% of a crop comprises discolored and empty seeds; seed viability is highest during heavy seed crop years. Collect seeds by picking or stripping the strobiles while they are still green enough to hold together and placing them directly into bags. Spread strobiles out to dry until disintegration begins, then shatter them by flailing and shaking and separate the

References

Brinkman, K.A. 1974. *Betula* L. Birch. pp. 252-57 *In*: Schopmeyer, C.S. (tech. coord.) 1974. *Seeds of the Woody Plants in the United States*. Agric. Handbook 450. Washington, DC: USDA Forest Service. 883p.

Dirr, M.A., and C.W. Heuser. 1987. *The Reference Manual of Woody Plant Propagation: From Seed to Tissue Culture*. Athens, GA: Varsity Press, Inc. 239p.

Elias, T.S. 1980. *The Complete Trees of North America Field Guide and Natural History*. New York: Van Nostrand Reinhold Company. 948p.

Uchytil, R.J. 1991. *Betula papyrifera*. *In*: Fischer, William C. (comp.) The Fire Effects Information System [Monograph Online]. Missoula, MT: USDA Forest Service, Intermountain Fire Sciences Laboratory. http://www.fs.fed.us/database/feis/plants/Tree/BETPAP. Accessed June 25, 1996.

seeds by screening and fanning. Seeds can be stored with a 1-5% moisture content at room temperature for 18-24 months. Unstratified seed germinates better than stratified seed. Sow seed during late summer or fall, or in spring after four to eight weeks of cold stratification. First-year seedlings are 5-12 cm tall. Seedlings require moderate shade during the first summer and can be outplanted after one or two years (Brinkman 1974, Uchytil 1991).

Seeds per kilogram: ~1,344,795-9,082,895 (Brinkman 1974)

Vegetative: Paper birch sprouts following cutting or fire, typically from the stump base or root collar. Prolific sprouting is common in young trees, with some individuals producing up to a hundred sprouts. Sprout growth is rapid, sometimes up to 60 cm in the first growing season (Uchytil 1991). Timing is critical when taking birch cuttings. Cuttings should be taken when shoots are still active with the base of the cutting just becoming firm. If the terminal bud is visible, results are usually poor. Nodal cuttings, 15-20 cm long, with a long shallow wound are best. Dip the cuttings in an 8000 ppm IBA talc rooting hormone before planting in a peat:sand medium. If cuttings root early, transplant them into pots, otherwise do not disturbed them. Growth can be accelerated for both seedlings and cuttings by placing them under long, warm day conditions with good air circulation and adequate moisture and nutrition (Dirr and Heuser 1987).

Chrysolepis chrysophylla
(Castanopsis chrysophylla)
Golden chinkapin

Description
Golden chinkapin is a medium to large tree that grows 15-40 m in height with a broad, rounded crown. The leaves are evergreen, alternate, lanceolate, 5-15 cm long, with entire but slightly wavy margins. The leaf texture is leathery and the underside is covered with golden-colored hairs. The bark is a thick, dark reddish-brown with deep fissures dividing the broad rounded ridges. The male catkins are 5-6.5 cm long and are produced at or near the tips of the branchlets. The female catkins are short and found clustered at the base of the stalks bearing the male flowers. The fruit is a nut that matures in two seasons and is enclosed in a spiny bur growing on the previous year's branchlets. The nuts are eaten by bears, deer, small mammals, and rodents, and some birds. The wood is soft, brittle, and fine grained and is used mainly for wood heating and campfires (Elias 1980).

Habitat and Geographic Range
Golden chinkapin grows in the mountain slopes of the Pacific Coast region at 1000-2000 m and ranges from western Washington south into central California. It is found in valleys and sheltered ravines. Soil conditions range from dry rocky soils to deep rich soils. Golden chinkapin grows in association with western juniper, canyon oak, and scrub oak (Elias 1980).

Propagation
Seed: The fruit ripens during August and September. Hand pick the burs in late summer or early fall, after ripening but before they open. Spread them out to dry in a warm room and run through a fruit disintegrator or shaker to separate the nuts. Store the nuts in sealed containers at 5°C; they will retain their viability for two to five years. Seeds do not need to be cold stratified and can be planted directly into containers larger than 10 cm^3 and covered with 5 cm of soil (Jebb 1995). Survival after emergence can be problematic (Hubbard 1974).
Seeds per kilogram: ~1,825-2,425 (Hubbard 1974)

References
Elias, T.S. 1980. *The Complete Trees of North America Field Guide and Natural History.* New York: Van Nostrand Reinhold Company. 948p.

Hubbard, R.L. 1974. *Castanopsis* (D. Don) Spach. chinkapin. pp. 276-77 *In*: Schopmeyer, C.S. (tech. coord.) 1974. *Seeds of the Woody Plants in the United States.* Agric. Handbook 450. Washington, DC: USDA Forest Service. 883p.

Jebb, T. 1995. Horticulturalist, USDI Bureau of Land Management, C.A. Sprague Seed Orchard, Merlin, OR. Personal communication.

Kruckeberg, A.R. 1982. *Gardening with Native Plants of the Pacific Northwest.* Seattle, WA: University of Washington Press. 252p.

McMurray, N.E. 1989. *Chrysolepis chrysophylla*. *In*: Fischer, William C. (comp.) The Fire Effects Information System [Monograph Online]. Missoula, MT: USDA Forest Service, Intermountain Fire Sciences Laboratory. http://www.fs.fed.us/database/feis/plants/Tree/CHRCHR. Accessed September 5, 1996.

Vegetative: Golden chinkapin regenerates vegetatively from adventitious buds found on stumps or basal burls (McMurray 1989). Propagation by cuttings is difficult, even with mist and a rooting hormone. Layering, grafting, and budding have been shown to be effective means of propagation (Kruckeberg 1982).

Cornus nuttallii
Pacific dogwood

Description

Pacific dogwood is a small, deciduous tree that can obtain heights up to 30 m, but usually grows no more than 9-12 m. The bark is reddish-brown and smooth or can have thin plates near the base. The leaves are opposite, simple, ovate to obovate in shape with arcunate veins and wavy-toothed margins. The flowers are produced in dense heads with four to six white petallike bracts. The fruit is a bright red drupe produced in dense rounded heads of thirty to forty. The fruit is a main staple of black bears and many birds. Beavers eat the fruits, wood, and leaves, while deer browse the twigs and leaves. The wood is extremely hard and fine grained and is used in making cabinets and tool handles. Pacific dogwood is a popular ornamental. It has a high flood resistance and is therefore effective in streambank stabilization with areas containing deep and well-drained soils. Douglas-fir regeneration can be slowed by the water-soluble leachates from senescent leaves of the Pacific dogwood (Randall et al. 1994, Elias 1980, Griffith 1992).

Habitat and Geographic Range

Pacific dogwood is found in bottomlands, moist river soils, and along stream banks from near sea level to 1820 m in elevation. It prefers deep, well-drained soils high in nitrogen with a rich humus layer. It ranges from southwestern British Columbia south on the west of the Cascade and Sierra Nevada Mountains into southern California (Randall et al. 1994, Elias 1980, Griffith 1992).

Propagation

Seed: Pacific dogwood reaches sexual maturity at six to ten years of age. Large seed crops are produced at two-year intervals. Seed maturity and dispersal occur from September to October. Collect by shaking or stripping the branches. Sow the seeds directly in the fall without removing them from the fruit or clean them for storage by macerating in water and allowing the pulp to float off. Seeds can be stored in sealed containers at 3-5°C for two to four years. Scarify stored seeds in concentrated sulfuric acid for four hours, rinse, then stratifiy at 3°C

References

Brinkman, K.A. 1974. *Cornus* L. Dogwood. pp. 336-42 *In*: Schopmeyer, C.S. (tech. coord.) 1974. *Seeds of the Woody Plants in the United States*. Agric. Handbook 450. Washington, DC: USDA Forest Service. 883p.

Elias, T.S. 1980. *The Complete Trees of North America Field Guide and Natural History*. New York: Van Nostrand Reinhold Company. 948p.

Griffith, R.S. 1992. *Cornus nuttallii*. *In*: Fischer, William C. (comp.) The Fire Effects Information System [Monograph Online]. Missoula, MT: USDA Forest Service, Intermountain Fire Sciences Laboratory. http://www.fs.fed.us/database/feis/plants/Tree/CORNUT. Accessed June 25, 1996.

Randall, W.R., R.F. Keniston, D.N. Bever, and E.C. Jensen. 1994. *Manual of Oregon Trees and Shrubs*. Corvallis, OR: Oregon State University Bookstores. 305p.

for 90 days and sow in the fall. Cover seed beds with up to 1.3 cm of soil and mulch with sawdust over winter (Brinkman 1974, Griffith 1992).

Seeds per kilogram: ~8,815-13,450 (Brinkman 1974)

Vegetative: Pacific dogwood resprouts from the root crown after disturbance by fire or logging. Take cuttings in June and July and treat with a rooting hormone. Cultivate in clay pots to help retain water and prevent root rot. Cuttings should be kept for no more than two years since transplants after this time have a high mortality rate. When transplanting, place Pacific dogwood in a ring of native shrubs to protect the lower trunk and branches from sunburn (Griffith 1992).

Juniperus communis
Mountain juniper

Description

Mountain juniper, also called common juniper, is a shrub or small tree that reaches 2-7 m tall and can have an open, irregular-shaped crown or can form a large 2-4 m-wide mat. The bark is dark reddish-brown and separates into long papery scales. Leaves are small, needle shaped, sharp pointed, with a broad white band on the upper surface, and spread in whorls of three at right angles to the branchlets. Male and female cones are commonly on separate trees. Male cones occur singly between the leaves and branchlets and are 4-5 mm long. Female cones occur singly in the junctions of the leaves near the end of the branchlets and are composed of three or four fleshy scales. The fruit is a berrylike cone, 5-8 mm in diameter and matures in two or three growing seasons, ripening to a glaucous bluish-black. There are one to three seeds per fruit. Cones are eaten by birds, squirrels, chipmunks, and raccoons. Mountain juniper is often planted as an ornamental (Elias 1980).

Habitat and Geographic Range

Mountain juniper often grows in disturbed soils, and grows well in poor, rocky, or gravelly soils. It can be found from sea level to 3400 m in elevation (Elias 1980) and ranges along the Rockies from British Columbia to New Mexico and Arizona and in the Cascades to northern California (Harlow et al. 1979). It can also be found in Europe, northern Asia, and Japan.

Propagation

Seed: Fruit ripens from August to October; collect by hand picking, stripping, or shaking the fruit from the tree onto canvas. Remove debris by fanning and extract seeds by macerating and floating away the pulp and empty seeds. Dry the seeds and plant or store at -5°C. They remain viable in storage for many years. Warm stratification with diurnal temperatures of 20°C (night) and 30°C (day) for 60-90 days followed by cold stratification for 90 days is required to induce germination. Sow seed in the fall or spring and cover with a thin layer of firm soil or sand. Seedlings are usually planted as three-year-old stock (Johnsen and Alexander 1974).

References

Elias, T.S. 1980. *The Complete Trees of North America Field Guide and Natural History.* New York: Van Nostrand Reinhold Company. 948p.

Harlow, W.M., E.S. Harrar, and F.M. White. 1979. *Textbook of Dendrology Covering the Important Forest Trees of the United States and Canada.* Sixth edition. New York: McGraw-Hill Book Company. 510p.

Snyder, L.C. 1991. Native Plants for Northern Gardens. Anderson Horticultural Library, University of Minnesota. 277p.

Johnsen, T.N. Jr., and R.A. Alexander. 1974. *Juniperus* L. Juniper. pp. 460-69 *In:* Schopmeyer, C.S. (tech. coord.) 1974. *Seeds of the Woody Plants in the United States.* Agric. Handbook 450. Washington, DC: USDA Forest Service. 883p.

Seeds per kilogram: ~56,105-120,150 (Johnsen and Alexander 1974)

Vegetative: Mountain juniper can be propagated by cuttings and will root in a sand medium (Snyder 1991).

Larix occidentalis
Western larch

Description

Western larch is a deciduous conifer reaching heights of 50-80 m and diameters of 140 cm. The trees may live to be over seven hundred years old. The bark is thick and has large plates with reddish-brown flakes. The thin, light green needles are 2.5-5 cm long, linear, flattened to triangular in cross section, and occur in whorls of fifteen to thirty at the tips of short spur shoots (Crane 1990, Randall et al. 1994). Male and female flowers are borne separately on the same tree. Cones are 2-4 cm long and are green-brown-purple when ripe. Cones have papery scales with small, pointed bracts extending beyond each scale. Western larch is used for construction lumber, utility poles, posts, and railroad ties because of its strength, hardness, and resistance to rot (Rudlof 1974, Crane 1990). Western larch needles are an important source of food for grouse. The seeds are occasionally eaten by birds and small rodents. Because of its deciduous needles, rapid early growth, and fire resistance, this species is useful for rehabilitation on well-drained moist sites within its range (Crane 1990). Native Americans had many uses for western larch, including a sweet syrup made from the sap, chewing gum from the pitch, and medicinal teas from the bark.

Habitat and Geographic Range

Western larch can be found from western Montana to eastern Oregon and Washington and southern British Columbia (Rudolf 1974). It tends to grow on moist sites at elevations ranging from 650 to 2450 m. It prefers deep, well-drained and fairly nutrient-rich soils, and appears to need calcium and magnesium. Soil types range from glacial till to volcanic ash (Crane 1990). It is shade intolerant and very demanding of moisture (Randall et al. 1994).

Propagation

Seed: Flowering occurs from April to June and seed dispersal is from September to October. Though it is usually a prolific seed producer, western larch can go up to fifteen years between good cone crops (Helliwell 1987). Pick larch cones from the tree in the fall as soon as they ripen and spread them out in thin layers to dry.

References

Crane, M.F. 1990. *Larix occidentalis. In*: Fischer, William C. (comp.) The Fire Effects Information System [Monograph Online]. Missoula, MT: USDA Forest Service, Intermountain Fire Sciences Laboratory. http://www.fs.fed.us/database/feis/plants/Tree/LAROCC. Accessed July 10, 1997.

Helliwell, R. 1987. Forest Plants of the Warm Springs Indian Reservation. Warm Springs, OR: Confederated Tribes of the Warm Springs. 177p.

Randall, W.R., R.F. Keniston, D.N. Bever, and E.C. Jensen. 1994. *Manual of Oregon Trees and Shrubs.* Corvallis, OR: Oregon State University Bookstores. 305p.

Rudolf, P.O. 1974. *Larix* Mill. Larch. pp. 478-85 *In*: Schopmeyer, C.S. (tech. coord.) 1974. *Seeds of the Woody Plants in the United States.* Agric. Handbook 450. Washington, DC: USDA Forest Service. 883p.

Open them by solar heat, by heating in a kiln (45°C for eight hours), by placing them in a heated room, or by breaking them up mechanically. After opening, shake to remove the seed, then dewing mechanically or by hand rubbing. Finally, clean the seed with a blower or fanning mill. Larch seed can be stored for three or more years if kept cold and dry in sealed containers. Larch germinates well without pretreatment. However, cool, moist stratification for 18 days may enhance germination of spring-sown seed (Rudlof 1974). Sow in the fall or spring and cover with not more than 0.5 cm of soil. Cover fall-sown seed with mulch. Outplant seedlings after one year (Rudolf 1974).

Seeds per kilogram: ~216,045-434,305 (Rudolf 1974)

Picea sitchensis
Sitka spruce

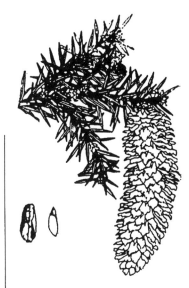

Description

Sitka spruce is an evergreen tree which can live to be over eight hundred years old. It is the largest spruce in the world, reaching heights of over 65 m and diameters of 5 m (Griffith 1992). The bark is thin, usually less than 2.5 cm thick, brown to gray in color, and scaly (Randall et al. 1994). The needles are 1.3-3 cm long, linear, with a white stomatal bloom on one side, and an attenuate apex that hurts when touched. The needles are flat to somewhat diamond shape in cross section and usually grow perpendicular to the twig (Randall et al. 1994). Male and female cones arise on separate branches of the same tree. Male strobili are ovoid to cylindrical shaped and pendant and are distributed throughout the crown. Female strobili are yellowish-green to purple in color, 0.5 to 1 cm in diameter, and arise at the end of branches in the top portion of the tree (Safford 1974). Sitka spruce is an important timber species due to its high strength to weight ratio. The wood is used for ladders, sail boat masts, saw timber, pulp, plywood, pianos, and guitars. Sitka spruce provides cover for a wide variety of wildlife and is browsed by some game birds (Griffith 1992). The native Makah people ate the young raw twigs as a source of vitamin C. Other native people used the pitch as medicine for burns, colds, sore throats, and toothaches, and the roots in making water-tight baskets (Pojar and MacKinnon 1994).

Habitat and Geographic Range

Sitka spruce can be found in a narrow strip along the northern Pacific coast from south-central Alaska to northern California (Griffith 1992). It grows from sea level to 600 m and is limited to areas with high precipitation and cool, moist summers. It grows best on deep, moist, well-drained soils but can tolerate salty ocean spray. It prefers soils high in calcium, magnesium, and phosphorus (Griffith 1992). Sitka spruce is a shade-intolerant species.

Propagation

Seed: Sitka spruce flowers in May and cones ripen from mid-August to mid-September. Collect cones at ripening

References

Griffith, R.S. 1992. *Picea sitchensis. In*: Fischer, William C. (comp.) The Fire Effects Information System [Monograph Online]. Missoula, MT: USDA Forest Service, Intermountain Fire Sciences Laboratory. http://www.fs.fed.us/database/feis/plants/Tree/PICSIT. Accessed July, 10 1997.

Pojar, J., and A. MacKinnon. 1994. *Plants of the Pacific Northwest Coast: Washington, Oregon, British Columbia, and Alaska.* Vancouver, BC, Canada: British Columbia Ministry of Forests and Lone Pine Publishing. 527p.

Randall, W.R., R.F. Keniston, D.N. Bever, and E.C. Jensen. 1994. *Manual of Oregon Trees and Shrubs.* Corvallis, OR: Oregon State University Bookstores. 305p.

Safford, L.O. 1974. *Picea A. Dietr. Spruce.* pp. 587-97 *In*: Schopmeyer, C.S. (tech. coord.) 1974. *Seeds of the Woody Plants in the United States.* Agric. Handbook 450. Washington, DC: USDA Forest Service. 883p.

to avoid seed loss. Air dry for several weeks or kiln dry for 6-24 hours at 40°C (Safford 1974). Extract the seeds by shaking or tumbling cones. Seeds can be stored for several years at 0-2°C in sealed containers. They do not require stratification but will germinate more uniformly following a cold, moist stratification period of 30 days. Sow to a depth of 0.5 cm in the spring. A thin layer of mulch is recommended (Safford 1974).

Seeds per kilogram: ~341,710-881,835 (Safford 1974)

Pinus albicaulis
Whitebark pine

Description

Whitebark pine is a long-lived, slow-growing tree (often reaching four to seven hundred years of age), growing 60-90 cm in diameter and reaching only 2-15 m in height. It usually has a broad, irregularly shaped crown and a short trunk. The bark is smooth and brownish-white, turning dark brown and separating into fissures upon aging. The needles occur in bundles of five, are 2.5-6 cm long, stiff, and clustered toward the end of the branchlets. The male and female cones are 8-10 mm long. The male cones are oval shaped and scarlet in color while the female cones are oblong and are a brighter scarlet color. The fruit is an ovoid-shaped cone, 2.5-7.5 cm long, with a stiff pointed tip on each scale. The seeds are wingless, somewhat flattened, broadest near the center and taper to a point. The seeds are eaten by squirrels, grizzly bears, and other wildlife. Whitebark pine has limited commercial value but is beneficial for watershed protection and aesthetics (Elias 1980).

Habitat and Geographic Range

Whitebark pine grows in cold, windy, snowy, and generally moist climatic zones at elevations of 2350-3750 m from northern British Columbia to southern California. It can grow on rocky ridges and bluffs but grows largest at lower elevations in protected ravines and canyons. It commonly grows on immature soils and can be found in association with Engelmann spruce and lodgepole pine (Elias 1980, Arno and Hoff 1989).

Propagation

Seed: Whitebark pine has enormously irregular cone crops. The cones ripen in August and September; collect them when they turn from dark purple to a dull purple to brown by hand picking. They should be collected as soon as they are ripe since most can be harvested and cached in a very short time by birds, chipmunks, and squirrels. Cones take 15-30 days to dry and open; dry them immediately after collection by spreading them out on a dry surface in the sun, or on trays in a well-ventilated building. Release seeds by breaking up the cones. Dry the seeds to a moisture content of 5-10% and

References

Arno, S.F., and R.J. Hoff. 1989. Silvics of whitebark pine (*Pinus albicaulis*). Ogden, UT: USDA Forest Service, Intermountain Research Station. Gen. Tech. Rep. INT-253. 11p.

Elias, T.S. 1980. *The Complete Trees of North America Field Guide and Natural History.* New York: Van Nostrand Reinhold Company. 948p.

Krugman, S.L. and J.L. Jenkinson. 1974. *Pinus L. Pine.* pp. 598-638 *In:* Schopmeyer, C.S. (tech. coord.) 1974. *Seeds of the Woody Plants in the United States.* Agric. Handbook 450. Washington, DC: USDA Forest Service. 883p.

Lueck, D. 1980. Ecology of *Pinus albicaulis* on Bachelor Butte, Oregon. M.A. thesis, Oregon State University. Corvallis, OR. 90p.

keep at -17 to -15°C for long-term storage and 1-5°C for short-term storage. Stratify by soaking the seeds in water for one or two days, then placing them in a moist medium and keeping them at a temperature of 1-5°C for 90-120 days. Whitebark pine generally has a poor germination rate (<35%). Germination can be enhanced by making a small cut in the seed coat to facilitate water uptake. Sow in late fall or early spring to a depth of 1.3 cm directly in nursery beds with well-aerated, fertile soil that allows for adequate drainage. Outplant as two- or three-year-old stock (Krugman and Jenkinson 1974, Lueck 1980, Arno and Hoff 1989).

Seeds per kilogram: ~4,850-6,615 (Krugman and Jenkinson 1974)

Vegetative: Cuttings should not be taken from trees older than five years (Krugman and Jenkinson 1974). Whitebark pine is easily grafted on stock plants of either whitebark pine or western white pine. It spreads only to a minor extent through layering (Arno and Hoff 1989).

Pinus ponderosa
Ponderosa pine

Description

Ponderosa pine is a large evergreen tree commonly reaching heights of 30-40 m and diameters of over 75-125 cm. Trees live for three to six hundred years (Habeck 1992). The bark of mature trees has large scaly plates which fit together like puzzle pieces. The needles are 12-25 cm long and occur mainly in bundles of three, occasionally two (Randall et al. 1994). The fruit is a 7-15-cm-long cone. Ponderosa pine is a valuable commercial species. Old-growth trees produce high-grade lumber used for molding, mill work, cabinets, doors, and windows. Ponderosa pine provides cover for many wildlife species. In addition, the needles, cones, buds, pollen, twigs, and seeds provide food for many species of birds and mammals (Habeck 1992).

Habitat and Geographic Range

Ponderosa pine can be found from southern British Columbia south to central Idaho and through the Cascade Range, Coast Ranges, and Sierra Nevada to southern California (Krugman and Jenkinson 1974). It grows on warm, dry sites with a short growing season and little annual precipitation, with most occurring during the winter months as snow. Ponderosa pine grows best on wet, deep, sandy gravel and clay loam soils, but is most often found on loams, loamy sand, or gravel (Habeck 1992). Its elevational range is 100 to 2700 m.

Propagation

Seed: Ponderosa pine flowers from April to June and cones ripen from August to September. Large seed crops are produced every two to five years. Collect cones when they are ripe but before seeds are shed. Dry immediately after collection to avoid molding by spreading them out in a well-ventilated area. Store dried cones in well-ventilated bags or trays. Remove seeds by shaking or tumbling the cones. Clean by dewinging and fanning. Seeds will remain viable in cold storage for up to 20 years. Cold stratify stored seed for 30-40 days. Sow in fall or spring to a depth of 0.5 cm. With fall-sown seeds, sowing must be late enough to avoid fall

References

Habeck, R.J. 1992. *Pinus ponderosa var. ponderosa. In*: Fischer, William C. (comp.) The Fire Effects Information System [Monograph Online]. Missoula, MT: USDA Forest Service, Intermountain Fire Sciences Laboratory. http://www.fs.fed.us/database/feis/plants/Tree/PINPONP. Accessed July 10, 1997.

Krugman, S.L. and J.L. Jenkinson. 1974. *Pinus* L. Pine. pp. 598-638 *In*: Schopmeyer, C.S. (tech. coord.) 1974. *Seeds of the Woody Plants in the United States.* Agric. Handbook 450. Washington, DC: USDA Forest Service. 883p.

Randall, W.R., R.F. Keniston, D.N. Bever, and E.C. Jensen. 1994. *Manual of Oregon Trees and Shrubs.* Corvallis, OR: Oregon State University Bookstores. 305p.

germination. Pine seed are often susceptible to losses from rodents and birds (Krugman and Jenkinson 1974).
Seeds per kilogram: ~15,210-50,705 (Krugman and Jenkinson 1974)
Vegetative: Pine can be propagated by rooting or grafting. However, rooting success tends to decrease when scions are taken from trees older than five years (Krugman and Jenkinson 1974).

Populus tremuloides
Quaking aspen

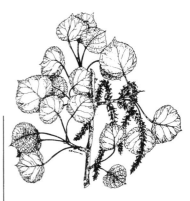

Description

Quaking aspen is a fast-growing tree that reaches an average height of 15-18 m and a diameter of 0.3-0.6 m. The leaves are simple, alternate, ovate shaped with serrate margins and have a flattened stalk which makes them appear to tremble in a breeze. The bark of young trees is smooth and whitish-green with rough dark scars along the bole. As the tree ages, the base turns a dark brown or gray and becomes furrowed. Twigs are slender and reddish-brown, with pointed buds. Aspen is dioecious, with flowers present on dense, drooping catkins 4-6 mm in length. The fruit is a two-valved, light green to brown capsule. Aspen provides browse for game animals and beaver. Many birds eat the buds, catkins, and seeds. The wood is light and weak, so is not very useful for furniture or construction, but is used for fences, barn doors, crates, boxes, and in making paper (Elias 1980, Harlow et al. 1979).

Habitat and Geographic Range

Quaking aspen can be found in a variety of sites from near sea level to 3050 m. Colder inland climates are preferred over humid coastal climates. The soils best suited for aspens are well drained, rich in calcium, and loamy. Areas also favorable to aspens are those where the soil is disturbed by fire and logging, and mineral soil is exposed. Quaking aspen is a very shade-intolerant species that tends to grow along forest edges and waterways, and in groves with other aspens (Elias 1980, Harlow et al. 1979).

Propagation

Seed: Regeneration is almost solely by root suckers, although the female clones do produce viable seeds. Aspen seed matures from late May to mid-June. It is best to collect the cottony, wind-borne seeds about a week before the capsules open. Use a long pruner to cut branches with the female catkins. Place the end of the branch in water with a constant temperature of 8-10°C, a high air temperature, and low relative humidity. When the capsules open, remove the seed using a suction device. Dry seed for three days at 24°C; it will remain

References

Burr, K.E. 1986. Greenhouse production of quaking aspen seedlings. pp. 31-37 *In*: Proceedings: Inter-mountain Nurseryman's Association Meeting: August 13-15, 1985, Fort Collins, CO. USDA Forest Service, Rocky Mountain Forest and Range Experiment Station. Gen. Tech. Rep. RM-125.

Campbell, R.B., Jr. 1984. Asexual vs. sexual propagation of quaking aspen. pp. 61-65 *In*: The challenge of producing native plants for the Intermountain area. Proceedings: Inter-mountain Nurseryman's Association Conference: August 8-11, 1983, Ogden, UT: USDA Forest Service Intermountain Forest and Range Experiment Station. Gen. Tech. Rep. INT-168. 96p.

Elias, T.S. 1980. *The Complete Trees of North America Field Guide and Natural History.* New York: Van Nostrand Reinhold Company. 948p.

Harlow, W.M., E.S. Harrar, and F.M. White. 1979. *Textbook of Dendrology Covering the Important Forest Trees of the United States and Canada*. Sixth edition. New York: McGraw-Hill Book Company. 510p.

Schreiner, E.J. 1974. *Populus* L. Poplar. pp. 645-55 *In*: Schopmeyer, C.S. (tech. coord.) 1974. *Seeds of the Woody Plants in the United States*. Agric. Handbook 450. Washington, DC: USDA Forest Service. 883p.

Starr, G.H. 1971. Propagation of aspen trees from lateral roots. *Journal of Forestry* 69(12):866-67.

viable for a year if stored at 5°C with a moisture content of 5-8%. Optimal germination occurs at temperatures of 15-25°C. Surface sow in a very moist seedbed and kept well watered (Schreiner 1974, Campbell 1984, Burr 1986). **Seeds per kilogram:** ~5,511,460-6,613,760 (Schreiner 1974)

Vegetative: Aspen reproduces readily via root suckers. The best time to collect the lateral roots for propagation is during the dormant stage: early spring, late summer, or autumn. Early spring is favored due to the ease of collection and because a greater number of shoots is produced. Roots 1-2 cm in diameter and 2.5 cm in length were found to be most effective in trial plantings. Plant root cuttings 1.3 cm deep in vermiculite and place in a greenhouse to sucker for six weeks. Cut the suckers and plant in a vermiculite:perlite mixture and mist until rooting occurs (about two to three weeks). Transplant again to a peat:vermiculite mixture and allow to grow. Keep temperatures between 15° and 25°C (Starr 1971, Campbell 1984).

Populus trichocarpa
Black cottonwood

Description

Black cottonwood is a deciduous tree and is the largest of the American poplars, growing up to 60 m in height and 1.5 m in diameter. The bark on young trees is a smooth yellowish-gray that furrows and turns to a darker grayish-brown as the tree matures. Leaves are alternate, ovate to ovate-lanceolate shaped, 7.5-10 cm long, and 5-6.5 cm wide. The leaves are dark green on top and silvery white underneath. The margin is finely serrated. The buds are 1.5-2 cm long, pointed and light orange-brown in color. Male catkins are 3.5-5 cm long and female catkins are 6-7.5 cm long. The fruit is a globose, three-celled pubescent capsule. Leaves and young shoots are often browsed by game mammals, and birds feed on buds, flowers, and seeds. Black cottonwood is a valuable timber tree in many parts of the West. The wood is used for furniture, boxes, and high-grade paper. Black cottonwood also provides food, cover, and shade for a variety of wildlife species (Elias 1980, Holifield 1990).

Habitat and Geographic Range

Black cottonwood grows in moist soils to rich humus from sea level to 2700 m. However, best growth is obtained at low elevations on deep river soils. It is found from southeastern Alaska and British Columbia to Washington and Oregon and the mountains of southern California and northern Baja California. It also occurs inland through southwestern Alberta, western North Dakota, Wyoming, Utah, and Nevada. Black cottonwood is shade intolerant and is commonly found in association with Douglas-fir, western white pine, hemlock, redcedar, alder, spruce, birch, and maple (Elias 1980, Holifield 1990).

Propagation

Seed: Seeds mature from late May to mid-July. A good time to collect is when a small number of the capsules are beginning to open. Collect the capsules in paper sacks and allow them to finish opening in a warm room. Clean the seeds with an air stream. Germination can occur very rapidly so it is important to either sow or

References

Buis, S. 1997. Owner, Sound Native Plants, Olympia, WA. Personal communication.

Dirr, M.A., and C.W. Heuser, Jr. 1987. *The Reference Manual of Woody Plant Propagation: From Seed to Tissue Culture.* Athens, GA: Varsity Press. 239p.

Elias, T.S. 1980. *The Complete Trees of North America Field Guide and Natural History.* New York: Van Nostrand Reinhold Company. 948p.

Galloway, G., and J. Worrall. 1979. Cladoptosis: a reproductive strategy in black cottonwood? *Canadian Journal of Forest Research* 9:122-25.

Haeussler, S., D. Coates, and J. Mather. 1990. Autecology of common plants in British Columbia: A literature review. British Columbia Ministry of Forests. FRDA Report 158. 272p.

Heilman, P.E., and G. Ekuan. 1979. Effect of planting stock length and spacing on growth of black cottonwood. *Forest Science* 25:439-43.

Holifield, J. 1990. *Populus trichocarpa. In*: Fischer, William C. (comp.) The Fire Effects Information System [Monograph Online]. Missoula, MT: USDA Forest Service, Intermountain Fire Sciences Laboratory. http://www.fs.fed.us/database/feis/plants/Tree/POPTRI. Accessed March 26, 1997.

Schreiner, E.J. 1974. *Populus* L. Poplar. pp. 645-55 *In*: Schopmeyer, C.S. (tech. coord.) 1974. *Seeds of the Woody Plants in the United States*. Agric. Handbook 450. Washington, DC: USDA Forest Service. 883p.

Wilson, M.G. 1996. Restoration ecologist. Portland, OR. Personal communication.

store the seeds immediately. Air dry for four days at a 5-8% moisture content and then store at 5°C (Schreiner 1974). Moist seedbeds are essential for high germination rates and seedling survival depends on continuously favorable conditions during the first month (Holifield 1990).

Vegetative: Stem cuttings are the preferred method to propagate black cottonwood. Take cuttings during November through March when the tree is dormant (Galloway and Worrall 1979). Select cuttings from healthy, moderately vigorous, one- to three-year-old wood on plants growing in full sunlight. Discard the tip portion (3-10 cm) since it is generally not suitable (Wilson 1996). The cuttings should be 1-2 cm in diameter and 25-50 cm long (Heilman and Ekuan 1979). The basal cut should be slanted and just below a dormant bud. The top cut should be straight and 1-3 cm above a dormant bud (Wilson 1996). Cold store cuttings in large drums with holes in the bottom and layer with moist sand or damp peat moss. Plant in a rich, well-drained soil medium (Dirr and Heuser 1987) so that just one bud is above ground (Wilson 1996). Keep cuttings well watered. Cuttings can also be used as live stakes (up to 3.5 cm in diameter and 1.2 m long) and stuck directly into the ground on site (Buis 1997). Black cottonwood naturally reproduces via stump sprouts after disturbance (Haeussler et al. 1990, Holifield 1990).

Prunus emarginata
Bitter cherry

Description

Bitter cherry is a small deciduous tree or shrub which grows up to 8 m tall. The bark is dark reddish-brown with large, widely spaced horizontal pores. Leaves are alternate, simple, elliptical to oval shaped, 3-8 cm long, and irregularly serrated at the margin. The white flowers are small, contain five petals, and are produced in clusters of five to twelve in the junctions of the upper leaves. The one-seeded fruit is a bright red, cherrylike berry that is 8-15 mm in diameter. Deer and elk browse the young leaves and twigs. Bears, squirrels, rabbits, rodents, and birds eat the fruits. The wood is brittle, soft, and close grained, and is sometimes used for furniture in the Northwest (Elias 1980).

Habitat and Geographic Range

Bitter cherry grows in moist, open-wooded areas, along streams, and in cut-over or burned-over areas. It ranges in elevation from sea level to 2400 m and occurs from the Pacific coast to the Cascade Range and from British Columbia south to southern California, Arizona, and southwestern New Mexico (Elias 1980).

Propagation

Seed: Fruit matures from July to September, when the berry turns bright red. Collect by hand stripping or by placing sheets underneath the tree and beating or shaking it. Clean the seeds by macerating in water and floating off the pulp. If seeds are sown immediately they do not need to be dried; if stored for weeks or months, they need to be surface dried only. For storage of a year or more, reduce the moisture content of the seed below the surface-dry condition and store between 1-5°C in sealed containers. Bitter cherry seeds have embryo dormancy and require a period of afterripening in the presence of moisture and oxygen for adequate germination. Seeds should be cold stratified at 5°C for 90-126 days in a sand and peat mixture prior to spring sowing. For fall sowing, it is important that seeds are sown early enough to allow for after-ripening before the ground freezes (Grisez 1974). Susceptibility to dieback and disease is increased when planted in poorly drained

References

Elias, T.S. 1980. *The Complete Trees of North America Field Guide and Natural History.* New York: Van Nostrand Reinhold Company. 948p.

Grisez, T.J. 1974. *Prunus* L. Cherry, peach, and plum. pp. 658-73 *In*: Schopmeyer, C.S. (tech. coord.) 1974. *Seeds of the Woody Plants in the United States.* Agric. Handbook 450. Washington, DC: USDA Forest Service. 883p.

Huxley, A., M. Griffiths, and M. Levy. 1992. *New Royal Horticultural Society Dictionary of Gardening.* Vol 3. London: Macmillan Press Ltd. p. 733.

soils (Huxley et al. 1992). Outplant seedlings in one to two years. (Grisez 1974)

Seeds per kilogram: ~9,080-19,380 (Grisez 1974)

Vegetative: Softwood cuttings taken during spring and early summer can be used for propagation. Treat with a rooting hormone and grow with mist and bottom heat (Huxley et al. 1992). In addition, bitter cherry reproduces vegetatively by root crown and root sprouts.

Prunus virginiana
Common chokecherry

Description

Common chokecherry is a deciduous shrub or small tree and grows up to 8 m in height and 20 cm in diameter. The crown is irregularly shaped and the trunk is often crooked. The bark is smooth to slightly fissured, and grayish-brown in color. The leaves are 5-10 cm long, alternate, simple, obovately shaped with fine, sharp serrations along the margin. The flowers, which appear in spring, are white with five rounded petals found in clusters 8-15 cm long. The one-seeded fruit is a drupe, 8-10 mm in diameter, and can be red, black, or yellow in color. The wood is not valuable due to its small diameter and crooked shape. However, deer browse the young twigs and leaves, and bears, grouse, quail, and many other birds eat the fruit. The chokecherry is a very hardy tree and, although usually classified as a weed, it can be useful in erosion control (Elias 1980).

Habitat and Geographic Range

Common chokecherry is broadly distributed and can be found from British Columbia down to southern California at elevations of 800-2500 m. It grows in moist soils and on open sites along roads, fences, and the edges of forests. Due to its shade intolerance, it is one of the first species to establish itself on cutover land. It grows in association with pin cherry, aspen, paper birch, northern red oak, and red maple (Elias 1980).

Propagation

Seed: The seeds of chokecherry mature in August and September. Hand pick or strip into a container or collect by placing canvas underneath the bush and beating it. Cleaning and soaking seeds can increase germination. Cleaned seeds can be stored just below surface dry conditions in sealed containers at 1°C for up to five years. Chokecherry has seed dormancy necessitating an after-ripening period in the presence of moisture and oxygen. Cool, moist stratification of 3-5°C for 120-160 days will result in more consistent germination. Plant seed in the fall or spring to a depth of 1.3 cm in a moist, well-drained soil (Grisez 1974, Mulligan and Munro 1981). Dieback and disease susceptibility increase on poorly drained soils (Huxley et al. 1992).

References

Elias, T.S. 1980. *The Complete Trees of North America Field Guide and Natural History.* New York: Van Nostrand Reinhold Company. 948p.

Grisez, T.J. 1974. *Prunus* L. Cherry, peach, and plum. pp. 658-73 *In*: Schopmeyer, C.S. (tech. coord.) 1974. *Seeds of the Woody Plants in the United States.* Agric. Handbook 450. Washington, DC: USDA Forest Service. 883p.

Huxley, A., M. Griffiths, and M. Levy. 1992. *New Royal Horticultural Society Dictionary of Gardening.* Vol 3. London: Macmillan Press Ltd. p. 733.

Mulligan, G.A., and D.B. Munro. 1981. The biology of Canadian weeds. 51. *Prunus virginiana* L. and *P. serotina* Ehrh. *Canadian Journal of Plant Science* 61: 977-92.

Seeds per kilogram: ~6,635-18,520 (Grisez 1974)
Vegetative: Common chokecherry reproduces primarily via rhizomes. This species can also be propagated by softwood cuttings taken during the spring and early summer. Treat cuttings with a rooting hormone and grow with mist and bottom heat (Huxley et al. 1992).

Pseudotsuga menziesii var. menziesii
Coastal Douglas-fir

Description

Coastal Douglas-fir is a fast-growing, long-lived, evergreen tree that typically reaches heights of over 75 m and diameters of 1-2 m. Young trees have smooth, gray bark which contains many resin blisters while mature trees have thick, corky bark (Uchytil 1991). The green needles are 2.5 cm long and are arranged spirally on the stems. Male strobili are 2 cm long and are borne abundantly throughout the crown. Female strobili develop on the end of the branches in the upper half of the tree. Female cones have characteristic three-lobed protruding bracts (Owston and Stein 1974). Coastal Douglas-fir produces more timber than any other tree species in North America (Uchytil 1991). The wood is used for a variety of purposes including lumber, plywood, posts, flooring, veneer, pulp, and furniture. Douglas-fir seeds are a primary food source for many birds and small mammals. The wood was used by many of the native people for spear handles, harpoon shafts and barbs, spoons, caskets, and halibut and cod hooks. The pitch was used as a medicinal salve for wounds and skin irritations, for sealing joints of harpoon heads and fishhooks, and for caulking canoes (Pojar and MacKinnon 1994).

Habitat and Geographic Range

Coastal Douglas-fir can be found in southwestern British Columbia through western Washington and Oregon to central coastal California and east into the Cascade and Sierra Nevada ranges (Owston and Stein 1974). It ranges in elevation from sea level to 2300 m. It grows best on well-aerated, deep soils but will tolerate a wide range of soil textures and parent materials. It is a dominant tree species in the Pacific Northwest, occurring in nearly all forest series (Uchytil 1991).

Propagation

Seed: Douglas-fir seed reaches maturity in August or early September. Collect the cones when they develop a brownish or purplish tinge; this can be as early as mid-August in warm, low-elevation areas and as late as

References

Owston, P.W., and W.I. Stein. 1974. P*seudotsuga* Carr. Douglas-fir. pp. 674-83 *In*: Schopmeyer, C.S. (tech. coord.) 1974. *Seeds of the Woody Plants in the United States.* Agric. Handbook 450. Washington, DC: USDA Forest Service. 883p.

Pojar, J., and A. MacKinnon. 1994. *Plants of the Pacific Northwest Coast: Washington, Oregon, British Columbia, and Alaska.* Vancouver, BC, Canada: British Columbia Ministry of Forests and Lone Pine Publishing. 527p.

Uchytil, R.J. 1991. *Pseudotsuga menziesii* var. *menziesii. In*: Fischer, William C. (comp.) The Fire Effects Information System [Monograph Online]. Missoula, MT: USDA Forest Service, Intermountain Fire Sciences Laboratory. http://www.fs.fed.us/ database/feis/plants/ Tree/PSEMENM. Accessed July 10, 1997.

October at higher elevations. In any given locality, the ideal collection time lasts only two or three weeks (Owston and Stein 1974). Dry the cones in well-ventilated sacks. They can be stored for three to four months under dry, well-ventilated conditions. Open the cones by air drying in warm, dry weather or by air drying for 8-21 days, then heating in a kiln at 35-40°C for 16-48 hours. Separate and clean seeds by tumbling the dried cones, screening out the seed, dewinging, and fanning or blowing. Seeds can be stored at 6-9% moisture content at -5 to 0°C for 20 years. Sow in the fall and allow to stratify naturally overwinter or stratify at 0-4°C for 30-40 days and sow in the spring. Sow in the ground or in small containers (Owston and Stein 1974). **Seeds per kilogram:** ~33,950-116,845 (Owston and Stein 1974)

Vegetative: Douglas-fir can be propagated from stem cuttings taken from trees up to 9-12 years old. Dip the cuttings in a rooting hormone (Owston and Stein 1974).

Quercus garryana
Oregon white oak

Description

Oregon white oak is a deciduous tree ranging from 12 to 20 m in height and 60-100 cm in diameter. The bark has scaly ridges separated by shallow, irregular furrows, and is gray-brown in color. The leaves are alternate, 7.5-10 cm long, 5-8 cm wide, oblong to obovate; the margin has five to seven rounded lobes separated by moderately deep sinuses. The fruit is an acorn found singly or in pairs, 2.5-3 cm long, ovoid to obovoid, with a shallow cup enclosing less than a third of the nut. The acorns are food for deer, squirrels, bears, pocket gophers, woodpeckers, and many small rodents. The wood is hard and heavy and is used for furniture, cabinets, and general construction (Elias 1980).

Habitat and Geographic Range

Oregon white oak grows on almost any kind of soil, but is commonly found on hot, dry, rocky slopes, and grows best in alluvial soils. It ranges from southern British Columbia to the central coast of California, usually at lower elevations though it can be found up to 1350 m (Elias 1980). Oregon white oak is drought resistant and flood tolerant.

Propagation

Seed: Oregon white oak produces a large acorn crop every three to four years. Collect acorns from September to November by shaking onto canvas sheets or picking off the ground and sowing immediately (Olson 1974). Remove defective acorns by flotation, but this only works on fresh-picked acorns (Jebb 1995). Fresh acorns germinate rapidly under warm, moist conditions. A one-week soak before fall sowing is recommended (Jebb 1995) although Buis (1996) reports excellent germination with an overnight soak. Plant acorns to a depth of 1.3 cm in seed beds or tall pots that provide ample room for root development. Suggested medium is a mixture of soil, peat, and sand. It is important to protect the acorns from mice prior to germination (Buis 1996). Leave the seedlings in the seed bed or pots until the following autumn, then transplant. Cut the roots frequently to promote the fibrous root system necessary for successful

References

Buis, S. 1996. Owner, Sound Native Plants, Olympia, WA. Personal communication.

Elias, T.S. 1980. *The Complete Trees of North America Field Guide and Natural History.* New York: Van Nostrand Reinhold Company. 948p.

Jebb, T. 1995. Horticulturalist, USDI Bureau of Land Management, C.A. Sprague Seed Orchard, Merlin, OR. Personal communication.

Niemiec, S.S., G.R. Ahrens, S. Willits, and D.E. Hibbs. 1995. Hardwoods of the Pacific Northwest. Corvallis, OR: Forest Research Laboratory, Oregon State University. Research contribution 8. 115p.

Olson, D.F., Jr. 1974.
Quercus L. Oak. pp. 692-703 *In*: Schopmeyer, C.S.
(tech. coord.) 1974. *Seeds
of the Woody Plants in the
United States.* Agric.
Handbook 450.
Washington, DC: USDA
Forest Service. 883p.

Sheat, W.G. 1948.
*Propagation of Trees,
Shrubs, and Conifers.*
London: Macmillan and
Co. 479p.

transplanting. When possible, sow acorns where they are intended to grow (Sheat 1948). Seedlings can take ten or more years to grow 1 m in height.
Seeds per kilogram: ~165-225 (Olson 1974)
Vegetative: Oregon white oak sprouts vigorously. Sprouts arise from dormant buds at the root collar and along the trunk (Niemiec et al. 1995).

Quercus kelloggii
California black oak

Description
California black oak is a deciduous tree that grows up to 10-28 m tall with a broad, rounded crown. The bark is smooth when young, developing thick, irregular, dark reddish-brown plates with age. Leaves are alternate, 10-18 cm long, 5-10 cm wide, with five to seven deep lobes and pointed tips. The fruit is an acorn found singly or in clusters of two or three, 2.5-4 cm long, cylinder shaped, with the cup covering one-fourth to one-third of the nut. The acorns are eaten by deer, birds, rodents, and bear, and both deer and elk browse the foliage. The wood is of little economic importance and is used mostly for firewood and fence posts (Elias 1980).

Habitat and Geographic Range
California black oak is found on well-drained, deep, sandy, and gravelly soils of canyon floors and lower to mid-mountain slopes from elevations of 450-3000 m. It grows in Oregon and California in the coastal ranges and the Sierra Nevada range (Elias 1980).

Propagation
Seed: Acorns mature in two years; collect from late September to early November (Niemiec et al. 1995) by shaking them onto canvas sheets or picking them promptly after they fall to prevent mold from destroying the cotyledon. Plant acorns immediately or place under cool, moist storage conditions of 15-16°C until planting the following spring. Stratification is required for 30-45 days at 1-5°C in moist sand and peat for spring sowing. Mulch fall beds with leaves or straw (Olson 1974). Cut roots frequently to promote the fibrous root system necessary for successful transplanting (Sheat 1948).
Seeds per kilogram: ~110-325 (Olson 1974)
Vegetative: California black oak has profuse basal sprouts after cutting (Niemiec et al. 1995).

References
Elias, T.S. 1980. *The Complete Trees of North America Field Guide and Natural History*. New York: Van Nostrand Reinhold Company. 948p.

Niemiec, S.S., G.R. Ahrens, S. Willits, and D.E. Hibbs. 1995. Hardwoods of the Pacific Northwest. Corvallis, OR: Forest Research Laboratory, Oregon State University. Research contribution 8. 115p.

Olson, D.F., Jr. 1974. *Quercus* L. Oak. pp. 692-703 *In*: Schopmeyer, C.S. (tech. coord.) 1974. *Seeds of the Woody Plants in the United States*. Agric. Handbook 450. Washington, DC: USDA Forest Service. 883p.

Sheat, W.G. 1948. *Propagation of Trees, Shrubs, and Conifers*. London: Macmillan and Co. 479p.

Rhus glabra
Smooth sumac

References

Brinkman, K.A. 1974. *Rhus* L. Sumac. pp. 715-19 *In*: Schopmeyer, C.S. (tech. coord.) 1974. *Seeds of the Woody Plants in the United States.* Agric. Handbook 450. Washington, DC: USDA Forest Service. 883p.

Elias, T.S. 1980. *The Complete Trees of North America Field Guide and Natural History.* New York: Van Nostrand Reinhold Company. 948p.

Tirmenstein, D. 1988. *Rhus glabra. In*: Fischer, William C. (comp.) The Fire Effects Information System [Monograph Online]. Missoula, MT: USDA Forest Service, Intermountain Fire Sciences Laboratory. http://www.fs.fed.us/database/feis/plants/Tree/RHUGLA. Accessed February 7, 1997.

Description

Smooth sumac is a deciduous shrub or small tree that grows up to 7 m in height. The bark is smooth, gray in color, with slight ridges developing with age. The leaves are alternate, pinnately compound with ten to thirty leaflets. The leaflets are lance shaped, 5-9.5 cm long, with sharply serrated margins. The main roots can grow to 2.5 m deep with many shallow laterals and rhizomes reaching depths of 7-30 cm. Male and female flowers are found on different trees in clusters at the ends of the branchlets. The male flowers are 15-25 cm long. The female flowers are 10-15 cm long. The flowers are yellow-green with five petals, five stamens in the male, and a single pistil in the female. The fruit is a dark red, berrylike drupe, 3-5 mm in diameter, produced in clusters at the end of the branchlets. The drupes are eaten by many different birds and mammals. In addition, deer browse the twigs and leaves and rabbits eat the bark and young twigs. Smooth sumac provides cover for many small birds and mammals. It is a pioneer species and is good at stabilizing soil and preventing erosion (Elias 1980, Tirmenstein 1988).

Habitat and Geographic Range

Except for an area in North Dakota and northeastern Montana, smooth sumac ranges from southern Quebec to British Columbia, throughout the United States, and into Mexico. It is often found near streams, along the edges of woodlands, prairies, canyons, and on abandoned farmlands. It can be found in a variety of soil conditions ranging from shallow to moderately deep, dry to moist, and in a variety of soil textures. Smooth sumac is relatively shade intolerant. It is considered a climax indicator in many shrub and grassland communities and is apparent in many early seral communities (USDA 1971, Elias 1980, Tirmenstein 1988).

Propagation

Seed: Smooth sumac flowers from May to July. The fruit ripens in September and October and can be picked by hand as soon as ripe until late in the year. If collected early, fruit may need additional drying by spreading it out in thin layers. Break the dried clusters down by beating in canvas sacks and screen or fan to remove debris. Remove damaged seed by placing in water; viable seed will sink. Seed can be stored in sealed containers at 0-5°C for two to ten years. Germination is inhibited by the hard, impervious hull and seed coat. Pretreatment consists of soaking in concentrated sulfuric acid for one to three hours. A constant temperature of 20°C or alternating warm and cool temperatures can promote germination. Continuous light also promotes good germination. Sow in the spring and cover with 2 cm of soil. Germination can be erratic or delayed resulting in seedling establishment over an extended period of time and variation in survival (Brinkman 1974, Tirmenstein 1988).

Seeds per kilogram: ~52,910-277,780 (Brinkman 1974)

Vegetative: Smooth sumac spreads vegetatively via long, shallow rhizomes (Tirmenstein 1988).

USDA Agricultural Research Service. 1971. *Common Weeds of the United States.* New York: Dover Publications, Inc. 463p.

Salix bebbiana
Bebb willow

References

Brinkman, K.A. 1974. *Salix* L. Willow. pp. 746-50 *In*: Schopmeyer, C.S. (tech. coord.) 1974. *Seeds of the Woody Plants in the United States.* Agric. Handbook 450. Washington, DC: USDA Forest Service. 883p.

Elias, T.S. 1980. *The Complete Trees of North America Field Guide and Natural History.* New York: Van Nostrand Reinhold Company. 948p.

Huxley, A., M. Griffiths, & M. Levy. 1992. *New Royal Horticultural Society Dictionary of Gardening.* Vol 4. London: Macmillan Press Ltd. p 163.

Jebb, T. 1995. Horticulturalist, USDI Bureau of Land Management, C.A. Sprague Seed Orchard, Merlin, OR. Personal communication.

Tesky, J.L. 1992. *Salix bebbiana. In*: Fischer, William C. (comp.) The Fire Effects Information System [Monograph Online]. Missoula, MT: USDA Forest Service, Intermountain Fire Sciences Laboratory. http://www.fs.fed.us/database/feis/plants/Tree/SALBEB. Accessed April 4, 1997.

Description

Bebb willow is a deciduous large bush or small tree that grows up to 7.5 m in height. The bark is smooth and gray and becomes furrowed as it ages. The leaves are alternate, simple, elliptical in shape, with evident meshed veins, and smooth margins. Male and female catkins are borne on separate trees. The fruit is a capsule, 5-9 mm long, that contains a number of cottony seeds. Deer, elk, moose, beaver, grouse, and hare browse young shoots, inner bark, and buds. The wood is used for furniture, baseball bats, and wickerwork (Elias 1980).

Habitat and Geographic Range

Bebb willow grows in moist, rich soils, such as streambanks and along lakes from 800 to 3000 m in elevation. It is distributed throughout much of the United States and Canada. It is a pioneer species and moves into moist, cleared areas (Elias 1980).

Propagation

Seed: Bebb willow flowers from April through July and the fruits ripen soon after flowering. Collect the seeds by picking from the tree or from drifts along the shore if found near water. It is unnecessary to separate the seed from the capsule. Seed is viable for only a few days, though if moistened and refrigerated in sealed containers, it can be stored for up to a month. Sow on beds and pack lightly with a roller. Keep seedbeds moist until seedlings have established. Shading with burlap or slats helps to conserve moisture and maintain a high relative humidity. Thin the seedbeds and trabsplant the seedlings after three to four weeks (Brinkman 1974).
Seeds per kilogram: ~5,510 (Brinkman 1974)
Vegetative: Take cuttings 30 cm long from one-year-old wood during late autumn or early spring. Make a slice cut at the base of the cutting. Plant cuttings into any soil type, but preferably a damp heavy soil (Huxley et al. 1992, Jebb 1995). Roots and shoots from cuttings will appear ten to twenty days after planting. Bebb willow will also reproduce vegetatively via root shoots, basal stem sprouting, and layering (Tesky 1992).

Salix lasiandra
Pacific willow

Description

Pacific willow is a deciduous tree that can reach 18 m in height. The grayish-brown bark becomes furrowed with broad, flat plates. The leaves are alternate, simple, narrowly elliptical, with finely toothed margins, and are green above while whitish below. Flowers are dioecious catkins. The male catkin ranges from 1.5 to 7 cm and the female from 2 to 7 cm. The fruit is a lance-shaped capsule 6-8 mm long. Deer and moose browse the foliage and twigs, while birds eat the catkins, young tips, and buds. It is heavily consumed by beaver during the winter months. The endangered Least Bell's Vireo (*Vireo bellii pusillus*) uses the Pacific willow to nest. Willows produce salicin which is closely related chemically to aspirin. The wood is very brittle and has no commercial use (Elias 1980, Uchytil 1989).

Habitat and Geographic Range

Pacific willow ranges from Alaska southeast to Saskatchewan and the Black Hills of South Dakota, southwest through the Rocky Mountains, to New Mexico and southern California. It can be found growing in well-drained sandy loams, to rich, rocky, or gravel soils from sea level to 2550 m. Pacific willow grows in association with other willows, red alder, and black cottonwood, usually along streambanks, lakes, and waterholes (Elias 1980, Uchytil 1989).

Propagation

Seed: Pacific willow flowers in the spring with the appearance of the leaves. Seeds are dispersed from spring through early summer. Sow in flats in a sand:perlite:vermiculite:peat (inorganics:organics; 4:1) medium and set in a shade house. After germination, transplant to individual containers and keep moist in a shade house for three to four months (Uchytil 1989, Evans 1992).
Seeds per kilogram: ~25,350 (Brinkman 1974)

References

Brinkman, K.A. 1974. *Salix* L. Willow. pp. 746-50 *In*: Schopmeyer, C.S. (tech. coord.) 1974. *Seeds of the Woody Plants in the United States.* Agric. Handbook 450. Washington, DC: USDA Forest Service. 883p.

Darris, D.C., T.R. Flessner, and J.D.C. Trindle. 1994. Corvallis Plant Materials Center Technical Report: Plant Materials for Streambank Stabilization, 1980-1992. Portland, OR: USDA Natural Resources Conservation Service. 172p.

Edson, J.L., A.D. Leege-Brusven, and D.L. Wenny. 1995. Improved vegetative propagation of Scouler willow. *Tree Planters' Notes* 46(2):58-63.

Elias, T.S. 1980. *The Complete Trees of North America Field Guide and Natural History.* New York: Van Nostrand Reinhold Company. 948p.

Evans, J.M. 1992. Propagation of riparian species in southern California. pp. 87-90 *In*: Landis, T.D. (tech. coor.), Proceedings, Intermountain Forest Nursery Association. USDA Forest Service. Gen. Tech. Rep. RM-211.

McCluskey, D.C., J. Brown, D. Bornholdt, D.A. Duff, and A.H. Winward. 1983. Willow planting for riparian habitat improvement. USDI Bureau of Land Management. Tech. Note 363. 21p.

Platts, W.S., C. Armour, G.D. Booth, M. Bryant, J.L. Bufford, P. Cuplin, S. Jensen, G.W. Lienkaemper, G.W. Minshall, S.B. Monsen, R.L. Nelson, J.R. Sedell, and J.S. Tuhy. 1987. Methods for Evaluating Riparian Habitats with Applications to Management. USDA Forest Service. Gen. Tech. Rep. INT-221. 177p.

Uchytil, R.J. 1989. *Salix lasiandra*. *In*: Fischer, William C. (comp.) The Fire Effects Information System [Monograph Online]. Missoula, MT: USDA Forest Service, Intermountain Fire Sciences Laboratory. http://www.fs.fed.us/database/feis/plants/Tree/SALLAS. Accessed January 30, 1997.

Vegetative: Take cuttings from mid-fall to early spring (when the leaves are off) from one- to four-year-old wood. Stems should be greater than 1 cm in diameter and cut to lengths of 30-50 cm. The terminal end should have a horizontal cut while the basal end should be cut at a 45° angle. Bundle the stems into groups of fifty or a hundred and dip into a fungicide. Cuttings can be stored for long periods by wrapping them in plastic bags and keeping them just above freezing. For short periods of storage, wrap them in moist burlap, sawdust, peat, or newspaper, and keep out of direct sunlight. Preroot cuttings in a greenhouse and allow to harden off prior to outplanting. They can also be planted directly on site. Clearings of 50-75 cm around the cutting help reduce competition. Plant with 25-40% of the cutting left above ground (McCluskey et al. 1983, Platts et al. 1987).

Disease: Willows have a few pests that cause problems with newly established plantings both in the nursery and on site. Many cuttings die on the rooting bench with stem canker symptoms typical of *Cytospora* spp. The poplar-and-willow borer (*Cryptorhynchus lapathi*) can kill young trees or severely deform them. Another pest is the willow blight, which is actually two pathogens, the willow scab (*Venturia saliciperda*) and the black canker (*Glomerella miyabeana*), that cause similar infections. The pathogens work alone or together and attack current-year leaves and stems. This causes tip dieback and defoliation which severely injures or can kill young trees. The tissue becomes less susceptible as trees age (Darris et al. 1994, Edson et al. 1995).

Salix scouleriana
Scouler willow

Description
Scouler willow is a deciduous tree 7.5 m tall with a high narrow crown. The bark is divided into broad ridges and is dark reddish-brown. Leaves are alternate, simple, obovate to elliptical, short-pointed at the tip, with smooth to wavy margins, and are covered with reddish hairs underneath. Scouler willow is a dioecious plant. The fruit is a capsule with numerous hairs. The staminate catkins are 2-4 cm long and the pistillate catkins are 2-6 cm long. Scouler willow is an important winter browse for large game animals. The shoots, buds, and catkins are eaten by rodents and other small mammals. The wood has no commercial value (Elias 1980, Uchytil 1989).

Habitat and Geographic Range
Scouler willow ranges from Alaska south along the coast to southern California and New Mexico, and northeast to the Black Hills of South Dakota. It is found from sea level to 3000 m on both moist lowland and dry upland soils. Scouler willow is a shade-intolerant plant and can be found growing in meadows and bogs, along rivers and roadsides, and in clearcuts (Elias 1980, Uchytil 1989).

Propagation
Seed: Plants flower from April to July and seed is dispersed soon after ripening. Seeds are nondormant and without moisture remain viable for only a few days. Seed can be stored up to six weeks at 0-5°C if kept moist. Light is required for good germination since Scouler willow seeds contain significant amounts of chlorophyll. Germination is normally high (95-100%) and usually occurs within two to five days (Uchytil 1989, King County 1994).
Seeds per kilogram: ~14,330 (Brinkman 1974)

References
Brinkman, K.A. 1974. *Salix* L. Willow. pp. 746-50 *In*: Schopmeyer, C.S. (tech. coord.) 1974. *Seeds of the Woody Plants in the United States.* Agric. Handbook 450. Washington, DC: USDA Forest Service. 883p.

Darris, D.C., T.R. Flessner, and J.D.C. Trindle. 1994. Corvallis Plant Materials Center Technical Report: Plant Materials for Streambank Stabilization, 1980-1992. Portland, OR: USDA Natural Resources Conservation Service. 172p.

Edson, J.L., A.D. Leege-Brusven, and D.L. Wenny. 1995. Improved vegetative propagation of Scouler willow. *Tree Planters' Notes* 46(2):58-63.

Elias, T.S. 1980. *The Complete Trees of North America Field Guide and Natural History.* New York: Van Nostrand Reinhold Company. 948p.

King County Department of Public Works, Surface Water Management Division. 1994. Northwest Native Plants: Identification and Propagation for Revegetation and Restoration Projects. King County, WA. 68p.

McCluskey, D.C., J. Brown, D. Bornholdt, D.A. Duff, and A.H. Winward. 1983. Willow planting for riparian habitat improvement. USDI Bureau of Land Management. Tech. Note 363. 21p.

Uchytil, R.J. 1989. *Salix scouleriana. In*: Fischer, William C. (comp.) The Fire Effects Information System [Monograph Online]. Missoula, MT: USDA Forest Service, Intermountain Fire Sciences Laboratory. http://www.fs.fed.us/database/feis/plants/Tree/SALSCO. Accessed January 28, 1997.

Vegetative: Scouler willow can be propagated by both hardwood and softwood cuttings. For hardwood cuttings, harvest 1-m-long dormant whips in late winter to early spring. If storage is necessary, wrap cuttings in moist burlap, or in plastic bags with moist sawdust, peat, or newspaper and store at 2°C. Cut the whips into shorter lengths of 8-12 cm (30-60 cm for direct planting) with a 45° cut made directly below a node at the basal end and a horizontal cut made at the terminal end. Treat the basal ends with a combination of growth hormone and fungicide. Stick cuttings in a mixture of peat, perlite, and vermiculite (1:1:1) and place on a rooting bench. Once rooted, transplant to larger containers and fertilize. Collect softwood cuttings in mid-summer. Remove shoot tips and lower leaves, but leave the uppermost leaves on the stem. From here, treat like hardwood cuttings (McCluskey et al. 1983, Edson at al. 1995).

Disease: Willows have a few pests that cause problems with newly established plantings both in the nursery and on site. Many cuttings die on the rooting bench with stem canker symptoms typical of *Cytospora* spp. The poplar-and-willow borer (*Cryptorhynchus lapathi*) can kill young trees or severely deform them. Another pest is the willow blight, which is actually two pathogens, the willow scab (*Venturia saliciperda*) and the black canker (*Glomerella miyabeana*), that cause similar infections. The pathogens work alone or together and attack current-year leaves and stems. This causes tip dieback and defoliation which severely injures or can kill young trees. The tissue becomes less susceptible as trees age (Darris et al. 1994, Edson et al. 1995).

Thuja plicata
Western redcedar

Description

Western redcedar is a large evergreen tree reaching heights of 20-35 m and diameters of 0.5-1.5 m. The bole is buttressed at the base and the crown is conical. The branches curve upward at the ends. The fibrous bark is thin and shreddy. The flattened, scalelike leaves are 1.5-3 mm long. Male and female flowers are borne on the same tree but usually on separate branches. The cones are about 1.3 cm long, erect, and ovoid to cylindrical in shape. The cone scales are semi-woody and thin, and number ten to twelve of which only about six are fertile, and with a small reflexed spine near the apex (Randall et al. 1994). Although its wood is soft and low in strength, western redcedar is a valuable timber species because of its light weight and resistance to decay. The wood is used for shingles, shakes, exterior siding, boats, caskets, and clothes closets. The leaf oil is used to make perfumes, insecticides, veterinary soaps, shoe polishes, and deodorants (Tesky 1992). Deer and elk browse the foliage and bears feed on the sapwood. Mature trees provide cover for many wildlife species. Western redcedar is held with the highest respect by native Northwest coast people and is called the "tree of life" by the Kwakwaka'wakw. The tree was used to make canoes, house planks, totem poles, paddles, baskets, clothing, dishes, ceremonial drum logs, and a variety of tools (Pojar and MacKinnon 1994).

Habitat and Geographic Range

Western redcedar can be found in the Pacific coast region from southeastern Alaska to northern California, the Cascade Mountains of Washington and Oregon, the Rocky Mountains in British Columbia, northern Idaho, and western Montana (Schopmeyer 1974). Western redcedar is very shade tolerant and commonly occurs as a dominant or codominant forest species on low-elevation (sea level to 2000 m) moist sites with cool summers and wet, mild winters. It grows on a wide range of soil types (Tesky 1992).

References

Pojar, J., and A. MacKinnon. 1994. *Plants of the Pacific Northwest Coast: Washington, Oregon, British Columbia, and Alaska.* Vancouver, BC, Canada: British Columbia Ministry of Forests and Lone Pine Publishing. 527p.

Randall, W.R., R.F. Keniston, D.N. Bever, and E.C. Jensen. 1994. *Manual of Oregon Trees and Shrubs.* Corvallis, OR: Oregon State University Bookstores. 305p.

Schopmeyer, C.S. 1974. *Thuja* L. Arborvitae. pp. 805-9 *In*: Schopmeyer, C.S. (tech. coord.) 1974. *Seeds of the Woody Plants in the United States.* Agric. Handbook 450. Washington, DC: USDA Forest Service. 883p.

Tesky, J.L. 1992. *Thuja plicata. In*: Fischer, William C. (comp.) The Fire Effects Information System [Monograph Online]. Missoula, MT: USDA Forest Service, Intermountain Fire Sciences Laboratory. http://www.fs.fed.us/database/feis/plants/Tree/THUPLI. Accessed July 10, 1997.

Propagation

Seed: Western redcedar flowers in late May to early June and cones ripen in early August. Collect cones when they have turned from yellow to brown. Collect from the tree or flail or strip onto a canvas or plastic sheet. Open the cones by drying in a 32°C kiln for 24-36 hours or at room temperature for a longer time. After drying, shake seeds out and separate from debris by fanning. Do not dewing the seeds (Schopmeyer 1974). Seeds can be stored for five or more years at -10 to 0°C. They germinate well without stratification. However, a cold, moist stratification at 1-3°C for 30-40 days can stimulate rapid germination of stored seeds. Sow in the spring to a depth of 0.5 cm. Seedlings require shade during their first year of growth (Schopmeyer 1974).

Seeds per kilogram: ~447,530-1,308,000 (Schopmeyer 1974)

Vegetative: Horticultural varieties of Thuja spp. are propagated from cuttings or by layering (Schopmeyer 1974). In closed canopies, western redcedar reproduces naturally from layering, and rooting of live branches or trunks which have fallen on wet soil (Tesky 1992).

Tsuga heterophylla
Western hemlock

Description

Western hemlock is a large evergreen tree commonly reaching heights of 30-45 m and diameter of 50-129 cm. The bark is rough, reddish-brown, thin in young trees, and becoming thick and furrowed in old trees (Pojar and MacKinnon 1994). The flat needles are 6-22 mm long and rounded at the tip. Male and female strobili develop in clusters at the end of branches. Mature cones are 2-3 cm long, oblong in shape, drooping, and sessil. The cone scales are thin and semi-woody (Randall et al. 1994). Western hemlock is one of the best pulpwood species for paper products. The wood is used for general construction, railroad ties, cabinets, boxes, and veneer for plywood (Tesky 1992). Deer and elk browse the foliage and mice consume the seed. Mature trees provide cover and habitat for many wildlife species. Western hemlock wood was used by western native people for spoons, combs, spearshafts, children's bows, and bowls. The bark was used in making dyes, and the boughs were used as bedding, skirts, and headdresses (Pojar and MacKinnon 1994).

Habitat and Geographic Range

Western hemlock can be found on the Pacific Coast from Alaska to northern California and in the mountains of northern Idaho and northwestern Montana (Ruth 1974). It is a shade species commonly occurring as a dominant or codominant on moist, low to mid-elevation (sea level to 2100 m) sites. It grows on all kinds of soil types and textures but grows better on well-aerated, nitrogen-rich soils (Tesky 1992).

Propagation

Seed: Western hemlock begins seed production when trees are twenty to thirty years of age. Flowering occurs from April to May and cones ripen from September to October. Because cones are small, they can be difficult to collect. Collect from the tops of trees felled for harvest or pick by climbing or using a pole pruner. Air dry cones in well ventilated bags for a few days up to a few months. Extract seeds by placing cones in a kiln for 48 hours, then tumbling or shaking the dry cones. Remove wings

References

Pojar, J., and A. MacKinnon. 1994. *Plants of the Pacific Northwest Coast: Washington, Oregon, British Columbia, and Alaska.* Vancouver, BC, Canada: British Columbia Ministry of Forests and Lone Pine Publishing. 527p.

Randall, W.R., R.F. Keniston, D.N. Bever, and E.C. Jensen. 1994. *Manual of Oregon Trees and Shrubs.* Corvallis, OR: Oregon State University Bookstores. 305p.

Ruth, R.H. 1974. *Tsuga* (Endl.) Carr. Hemlock. pp. 819-27 *In*: Schopmeyer, C.S. (tech. coord.) 1974. *Seeds of the Woody Plants in the United States.* Agric. Handbook 450. Washington, DC: USDA Forest Service. 883p.

Tesky, J.L. 1992. *Tsuga heterophylla. In*: Fischer, William C. (comp.) The Fire Effects Information System [Monograph Online]. Missoula, MT: USDA Forest Service, Intermountain Fire Sciences Laboratory. http://www.fs.fed.us/database/feis/plants/Tree/TSUHET. Accessed July 10, 1997.

and foreign matter in a fanning mill or gravity separator (Ruth 1974). Seeds can be stored at -5 to 0°C for five or more years. Soak seeds in cold water for 24-26 hours, then drain off excess water and cold stratify for 21-42 days prior to sowing. Surface sow in the spring. Seedlings need shade during their first year (Ruth 1974). **Seeds per kilogram:** ~416,665-1,119,930 (Ruth 1974) **Vegetative:** Western hemlock can be propagated by layering, cuttings, or grafting. Seedlings that die back to the soil surface will sprout from buds near the root collar (Tesky 1992).

Glossary

achene: a small, dry, indehiscent, one-seeded fruit, unwinged but often plumose

acorn: a nut topped with a scaly or bristly cap

acuminate: gradually tapering to a sharp point

achene

acute: terminating in a distinct but not protracted point, the converging edges separated by an angle less than 90°

adventitious: arising in abnormal positions, e.g. roots arising from the shoot system, buds arising elsewhere than in axils of leaves

aerobic: able to live or grow only where free oxygen is present

aggregate: collected together in tufts, groups, or bunches; applied especially to inflorescences and fruits

aggregate

allelopathy: suppression of germination, growth, or the limiting of the occurrence of plants as the result of the release of chemical inhibitors by some plants.

alternate: refers to leaf arrangement: single leaves alternate along the length of the twig

anaerobic: able to live and grow without air or free oxygen

angiosperm: a seed-bearing plant whose seeds develop within an enclosed ovary

annual: a plant that completes its life cycle in a single growing season

anther: the pollen-bearing portion of the stamen

anthesis: the flowering period, when the flower is fully expanded and functioning

apex: the tip

appressed: pressed closely against but not united with

arcuate: curved into an arch, like a bow

aristate: awned; provided with a bristle at the end or at the back or edge of an organ; in grasses, applies especially to the awns at the end of the bracts of the spikelet

articulate: jointed; joined by a line of demarcation between two parts which at maturity separate by a clean-cut scar

asexual: not forming part of a cycle which involves fertilization and meiosis

attenuate: tapering gradually to a narrow tip or base

auxins: hormones produced primarily in the plant shoot meristems that stimulate root initiation and development

awn: a slender bristle at the end or on the back or edge of an organ; in grasses, the awn is usually a continuation of the midnerve (and sometimes the lateral nerves) of the glumes or lemmas

axil: the angle between a leaf and the stem to which it attaches; *adj.* axillary

basal: relating to or situated at the base

basal sprouting: occurs when a plant sends up new shoots from buds located either at the base of the stem or on below-ground burls

basipetal: near the base rather than the tip; produced sequentially from the apex toward the base, as the flowers in a determinate inflorescence.

berry: a several-seeded indehiscent fruit, the outer and inner walls of which are fleshy, with the seeds embedded in the pulpy mass

biennial: a plant that completes its life cycle in two growing seasons

berry

bifid: deeply two-cleft or two-lobed, usually from the tip

biternate: having double groups of three leaflets

blade: the expanded portion of the leaf located above the sheath

bloom: a whitish, waxy, powdery coating on a surface; the flower

bract: a leaflike structure, different in form from the foliage leaves and without an axillary bud, associated with an inflorescence or flower

bud: small axillary or terminal structure on the stem or branch, consisting of embryonic foliage or floral leaves

bur: a rough or prickly propagule consisting of a seed or fruit and associated floral parts of bracts

callus: a protruding mass of hardened tissue, often formed after an injury but sometimes a regular feature of the plant; e.g. on the axis of the spikelet of some grasses

calyx: the outer perianth whorl; collective term for all of the sepals of a flower

cambium: a tissue composed of cells capable of active cell division, producing xylem to the inside of the plant and phloem to the outside; a lateral meristem

capsule: dry, dehiscent fruit; the product of a compound pistil splitting along two or more sutures

carpel: a simple pistil formed from one modified leaf, or that part of a compound pistil formed from one modified leaf

cartilaginous: tough and firm but elastic and flexible, like cartilage

capsule

catkin (ament): a long, narrow inflorescence composed of numerous flowers; typically unisexual (all female or all male); may be erect or pendulous; wind pollinated

caudex: the persistent and often woody base of a herbaceous perennial

catkin

chaff: thin, membranous scales or bracts; thin, dry unfertilized ovules among the fully developed seeds of a fruit

ciliate: with a marginal fringe of hairs

clone: a group of individuals originating from a single parent plant by vegetative reproduction

compound: refers to leaf composition; more than one leaf blade shares a common stem; the individual leaves are termed leaflets

cone: woody, leathery, or semi-fleshy scales spirally arranged, or alternating in pairs at right angles, and inserted on a central axis; each fertile scale bears one or more seeds; also called a strobile

cone

conifer: a plant that bears its seed inside woody or semi-woody strobiles (commonly called cones)

corymb: a racemose inflorescence in which the pedicels of the lower flowers are longer than those of the flowers above, bringing all flowers to about the same level

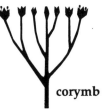

corymb

cotyledon: the primary leaf of the embryo; a seed leaf

crenate: refers to a leaf margin with rounded to blunt teeth

crown: the part of a tree or shrub above the level of the lowest branch

cryptogam: a plant that bears no flowers or seeds but propagates by means of spores, as algae, mosses, ferns, etc.

culm: The jointed stem of grasses, sedges, or rushes bearing the inflorescence

cuneate: wedge-shaped with the narrow part below

cutting: a vegetative portion of a plant which is separated from the parent plant and cultured to develop new roots and shoots; different types of cuttings are taken (leaf, nodal, basal, softwood, hardwood) based on the rooting capabilities of the parent species

cyme: a flat-topped or round-topped determinate inflorescence, paniculate, in which the terminal flower blooms first

damping-off: a fungal disease that attacks new seedlings as they first emerge from the soil; the condition is indicated when the top of the seedling suddenly wilts or leans over

deciduous: not persistent; the foliage falls seasonally

decompound: more than once-compound, the leaflets again divided

decumbent: reclining on the ground but with the tip ascending

dehiscent: refers to fruit which break open at maturity to release the contents (i.e. seeds)

deltoid: triangular

dentate: refers to a leaf margin with sharp teeth pointing outward

dentate

dewinger: a machine that removes the wing from the seed

dioecious: having the female and male reproductive structures on separate plants

disarticulating: separating at maturity at a joint

disk: in the Compositae (Asteraceae), the central portion of the involucrate head bearing tubular or disk flowers

disk flower: a regular flower of the Compositae (Asteraceae)

distal: toward the tip

diurnal: occurring or opening in the daytime; twenty-four-hour cycle

dormancy: general term for instances when a living tissue that is predisposed to grow does not do so; the seed coat can impose dormancy through its impermeability to water or gas exchange or mechanical

restrictions on growth of the embryo; agents or conditions within the mature seed can also maintain dormancy until specific environmental requirements are met

doubly serrate: refers to a leaf margin which is coarsely serrate with the teeth margins again serrated

drupe

drupe: a mostly one-seeded fleshy fruit, usually the product of a simple pistil, the outer wall fleshy, the inner wall bony

ecotype: physiological or morphological characteristics resulting from adaptation to specific environmental conditions which may be described by geographic location, climate, altitude, soil condition, etc.

entire: refers to a leaf margin which is smooth, without lobes or teeth

epicormic: of buds, shoots, or flowers, borne on the old wood of trees (applied especially to shoots arising from dormant buds after injury or fire)

entire

erose: with the margin irregularly toothed, as if gnawed

evergreen: bearing green leaves throughout the year

exfoliate: to peel off in flakes or layers, as the bark of some trees

exserted: projecting beyond the surrounding parts, as stamens protruding from a corolla; not included

fascicle: a tight bundle or cluster

filament: that part of the stamen which supports the anther

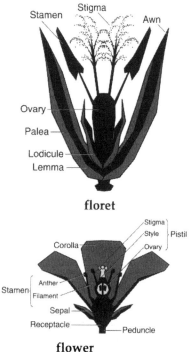

floret

flower

floret: a grass flower, together with the lemma and palea that enclose it

flower: the reproductive portion of the plant, consisting of stamens, pistils, or both

follicle: a dry, dehiscent, multiple-seeded fruit that splits open along one side at maturity

follicle

forage: the leaves and other plant parts eaten by herbivorous animals

fruit: the seed-bearing structure in angiosperms formed from the ovary after flowering

fusiform: spindle-shaped; broadest near the middle and tapering toward both ends

geniculate: with abrupt kneelike bends and joints

genotype: the total complement of hereditary factors (genes) acquired by an organism from its parents and available for transmission to its offspring

germination: the initial emergence of roots, shoots, and leaves from a seed usually in response to favorable

external conditions (e.g. temperature, moisture, and oxygen) following a period of dormancy

gibberellic acid: an acid used to increase the growth of plants and to improve fruit yield

glabrous: without hairs of any sort

globose: globe-shaped; spherical

glaucous: covered with a powdery or waxy coating

glume: a bract at the base of the spikelet or in the inflorescence of a grass, sedge, or similar plant

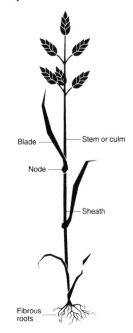

grass

grass: monocotyledonous herbaceous plants of the family *Poaceae* (also known as *Gramineae* family)

gymnosperm: a seed plant, usually a tree, which bears seeds not enclosed inside ovaries

habitat: the physical environment in which a plant or animal lives

hand-stripping: rubbing with the palm of the hand

hardwood cutting: a stem cutting taken from a deciduous plant after it has lost its leaves and before spring growth ensues; a stem cutting taken from a mature shoot of an evergreen plant during dormancy

head: densely packed cluster of stalkless flowers

heel-cutting: branchlet detached at juncture of main branch; the "heel" when trimmed forms callus tissue and roots in appropriate soil

herbaceous: not woody, applied to a plant whose above-ground parts do not form hard wood

hip: fruit consisting of the fleshy floral tube surrounding the mature ovaries, such as *Rosa* spp.

hispid: rough with firm, stiff hairs

hypocotyl: the portion of the embryonic stem below the cotyledons

IBA: indolebutryic acid, a synthetic auxin which is added to the base of cuttings to promote optimum rooting

imbibe: to absorb moisture

indehiscent: refers to dry fruits that normally do not split open at maturity

inflorescence: the group or arrangement in which flowers are borne on a plant

innovation: the basal shoot of a perennial grass

integument: the covering of the ovule which will become the seed coat

internode: the portion of a stem between two successive leaves or leaf pairs (or branches of an inflorescence)

involucre: a whorl of bracts subtending a flower or flower cluster

involute: refers to blades which are rolled inward from the edges, the upper surface within

keel: the sharp fold at the back of a compressed sheath, blade, glume, or lemma; the palea and sometimes the glumes and lemmas may be two-keeled

lammas: growth which develops in the late summer from the current season's bud

lanceolate: lance-shaped; several times longer than broad, widest at a point about one-third of the

lanceolate

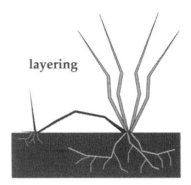

layering

distance from the base; narrowed at the ends

layering: a branch's production of roots when it comes into long-term contact with the soil or other rooting medium; layering branches typically have the ability to grow independently when separated from their parents

legume: dry, multi-seeded, dehiscent fruit; the product of a simple pistil splitting along two lines of suture

lemma: the outer or upper scalelike bract of a spikelet enclosing a grass flower

lenticels: round, oval or slitlike pores on the twigs, branches, and stems of a plant which are concerned with aeration

legume

ligneous: woody

ligule: the thin appendage or ring of hairs on the inside of a leaf at the junction of sheath and blade

linear: long and narrow with margins parallel or nearly so

lobed: refers to a leaf margin which is divided into lobes separated by rounded sinuses which extend from one-third to one-half of the distance between margin and midrib

maceration: a process for removing the soft, pulpy tissue from fleshy fruits

margin: the edge of a leaf or leaflet

meiosis: the two-stage division of a diploid nucleus, occurring once every sexual life cycle, in which gene recombination occurs and the number of chromosomes characteristic of the sporophyte plant is halved prior to the production of gametes

meristem: undifferentiated, actively dividing tissues at the growing tips of shoots and roots

micropyle: the opening in the integuments of the ovule

midrib: the central rib or vein of a leaf or other organ

monocotyledon: plants with a single seed leaf, or cotyledon

monoecious: having the female and male reproductive structures in separate flowers but on the same plant

morphology: the form and structure of an organism or part of an organism; the study of form and structure

mucronate: tipped with a short, sharp, abrupt point

mulch: any material such as straw, sawdust, leaves, or plastic film that is spread upon the surface of the soil to protect the soil and plant roots from the effects of rain, soil crusting, freezing, evaporation, etc.

mycorrhiza: a symbiotic union between a fungus and a plant root

nerve: the vascular veins (mostly longitudinal) of the blades, glumes, and lemmas

node: a joint on the stem at which leaves, bracts, or branches are produced

nut: a dry, one-seeded indehiscent fruit with a bony, woody, leathery, or papery wall, usually partially or wholly encased in a husk

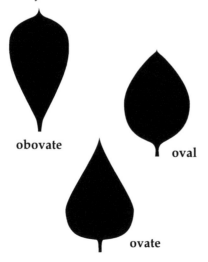

obovate

oval

ovate

oblanceolate: inversely lanceolate

oblong: longer than broad, and with sides nearly parallel

obovate: inversely ovate

obovoid: inversely ovoid, with the attachment at the narrower end

opposite: refers to leaf arrangement; single leaves are borne at the same height on the twig but are attached on opposite sides

oval: broadly elliptical, with the width greater than one-half the length

ovary: the basal portion of a flower's pistil that bears the ovules that develop into seeds

ovate: egg shaped, with the broadest part near the base

ovule: an immature seed; the megasporangium and surrounding integuments of a seed plant

palea: the inner or upper scalelike bract of a floret

palmately lobed margin

palmately compound

palmate: refers to compound leaf arrangement or leaf venation; all leaflets/veins arise from a common point

panicle: a compound raceme; an indeterminate inflorescence in which the flowers are borne on branches of the main axis or on further branches of these

pedicel: the stalk or stem of a spikelet; the opposite of sessile

panicle

peduncle: a primary flower stalk supporting either a cluster or a single flower

pendent: suspended or hanging, nodding

pendulous: hanging or drooping downward

perennate: maintain a dormant, vegetative state throughout non-growing seasons

perennial: a plant that lives for more than two growing seasons

perianth: the calyx and corolla of a flower, collectively, especially when they are similar in appearance

perigynium: a saclike bract enclosing the carpel in the genus *Carex*

perlite: an inorganic material used in growing media to promote drainage and porosity; an alumino-silicate mineral treated with heat and crushed to produce white, lightweight particles

persistent: remaining attached, either after other parts have been shed, or for a considerable period

petal: usually the showy, colored parts of a flower

petiole: the stem of a leaf or leaflet

phanerogam: a plant which produces seed

phenology: natural cycle based on season and climate

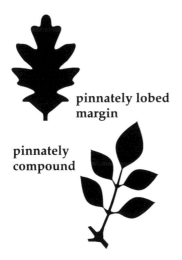

pinnately lobed margin

pinnately compound

phenotype: the physical characteristics of an organism; the outward expression of characteristics conferred on an organism by its genotype and influenced by its environment

pilose: pubescent with soft straight hairs

pinnate: refers to compound leaf arrangement or leaf venation; all leaflets/veins arise from a separate point

pioneer: a plant capable of invading newly exposed soil surfaces and persisting until supplanted by successor species

pistil: the female part of a flower, composed of ovary, stigma, and style

pith: the central region of a stem, inside the vascular cylinder

plagiotropism: occurs when a portion of a plant, such as a lateral branch, has a predisposition to grow non-vertical even when it becomes the main stem

plumose: feathery; with hairs or fine bristles on both sides of a main axis, as a plume

pollination: the transfer of pollen from the male organ, where it is formed, to the receptive region of the female organ, e.g. from anther to stigma

pome: fleshy fruit; the product of a compound pistil, the outer ovary wall fleshy, the wall papery or cartilaginous, and encasing numerous seeds

pome

primordia: aggregation of cells that will form a distinct organ

propagule: a structure with the capacity to give rise to a new plant, e.g. a seed, a spore, part of the vegetative body capable of independent growth if detached from the parent

proximal: toward the base or the end of the organ by which it is attached

puberulent: minutely pubescent; with fine, short hairs

pubescent: covered with short, soft hairs

raceme

pulvinus: a swelling or enlargement at the base of a petiole or petiolule

raceme: an inflorescence with stalked spikelets borne along a single axis

rachilla: a small rachis; applied especially to the axis of a spikelet

rachis: the axil of a spike or raceme; the stem of a compound leaf

radicle: pertaining to the root; arising from, or near, the roots

ray: the straplike portion of a ligulate flower (or the ligulate itself) in the Compositae (Asteraceae); a branch of an umbel ray flower: a ligulate flower of the Compositae

rhizome: an underground stem, usually growing horizontally, from which roots and shoots emerge along the internodes; usually persists from season to season

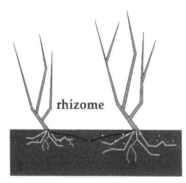

rhizome

riparian: growing along the banks of streams, rivers, or other bodies of water

root: the part of the plant below ground that takes up water and nutrients from the soil

rooting hormone: see **auxins**

root suckering: the sprouting of new shoots from stored or adventitious buds along a plant's root system when the plant is disturbed

rosette: a circular arrangement of leaves, usually at or near ground level

runner: an aboveground stem that creeps along the ground, and from which roots and shoots emerge at the nodes; also called stolon

samara: a dry, one-seeded, indehiscent fruit with a wing attached

samara

scaberulose: slightly rough to the touch, due to the structure of the apidermal cells, or to the presence of short stiff hairs

scabrous: covered with minute points, teeth, or very short stiff hairs which make it rough to the touch

scale: any thin, flat, scarious structure

scarification: the process of mechanically or chemically breaking the hard exterior coat of a seed in order to facilitate the penetration of water and atmospheric gases

scarious: thin, dry, and membranous in texture, not green

schizocarp: A dry, indehiscent fruit which splits into separate one-seeded segments (carpels) at maturity

seed: a propagating organ formed in the sexual reproductive cycle of gymnosperms and angiosperms, consisting of a protective coat enclosing an embryo and food reserves

sedge: grasslike, monocotyledonous plants, usually having triangluar, solid stems, three rows of narrow, pointed leaves, and minute flowers borne in spikelets

senescent: refers to aging and decay of a plant and its parts (e.g. leaves of deciduous plants)

sepal: a member of the outer whorl of nonfertile parts surrounding the fertile organs of a flower

serrate: refers to a leaf margin with sharp teeth

sessil: without a pedicel or stalk, the opposite of pediceled

serrate

sheath: The basal part of the leaf that wraps around the stem and holds it in place

shrub: a woody plant less than 5 m high either without a distinct main axis, or with branches persisting on the main axis almost to its base

simple

simple: refers to leaf composition; there is but one leaf blade on the petiole

sinus: the cleft, depression, or recess between two lobes of an expanded organ such as a leaf or petal

softwood cutting: a stem cutting taken from emerging plant shoots

spike: an unbranched inflorescence composed of stalkless flowers arranged on a single axis: also a general term for grass flowers

spikelet: an individual grass flower; most inflorescences are composed of many spikelets

spore: a reproductive cell resulting from meiotic cell division in a sporangium, representing the first cell of the gametophyte generation

stamen: the male organ of a flower, composed of anther and filament

sticking: placing a cutting into the propagation medium

stigma: that part of the pistil which receives the pollen

stolon: an aboveground stem that creeps along the ground, producing roots and new plant at the nodes; also called a runner; see **layering**

stomate: a pore or aperture, surrounded by two guard cells, which allows gaseous exchange

stratification: pregerminative treatment of seeds to break dormancy and to promote rapid uniform germination; accomplished by exposure to heat or cold, soaking, or other treatment of the seed

strobile: the reproductive structure of a gymnosperm (commonly called a cone); also, a dry, composite fruit in an angiosperm in which the individual fruits are achenes (e.g. alder)

strobile

style: that part of the pistil which connects the stigma with the ovary

sub: prefix meaning under, slightly, somewhat, or almost

subtend: to be below and close to, as a bract may subtend an inflorescence

subulate: narrow and tapering gradually to a fine point

succession: change in community composition and structure through time

suture: a line of fusion; the line of dehiscence of a fruit or anther

symbiosis: mutually beneficial relationship between two dissimilar living organisms. In some cases, the symbionts form a single body or organ, as in mycorrhizae or lichens

terete: round in cross section; cylindrical

terminal: at the end or top

ternate: in threes, as a leaf which is divided into three leaflets

tolerance: the ability of an organism to tolerate or withstand particular environmental conditions (e.g. shade, drought, etc.)

transpiration: emission of water vapor from the leaves, primarily through the stomata

tree: a woody plant at least 5 m high, with a main axis the lower part of which is usually unbranched

trifoliate: describes leaves divided into threes

triternate: triply ternate

umbel: an indeterminate inflorescence consisting of several pedicelled flowers all attached at the same point on the peduncle

umbel

vascular: specialized for conduction of fluids

vegetative: refers to the leaves, stems, and roots of the plant as contrasted with the reproductive parts, such as flowers or cones

vegetative propagation: asexual reproduction; the production of new individuals without genetic recombination ("parents" and "offspring" are genetically identical); some woody plants have the ability to produce new, genetically identical individuals through processes such as rhizome production, layering, basal sprouting, root suckering, and stolons; human-assisted techniques include grafting, rooted cuttings, and tissue culture

vermiculite: a lightweight, inorganic material with a high water- and nutrient-holding capacity used for cultivating plants; a heat-treated aluminum-iron-magnesium silicate mineral that consists of a series of thin, parallel plates

viability: capacity of a seed to germinate

villous: pubescent with long soft hairs

whorl: cyclic arrangement of appendages at a node

wing: a membrane expansion of a fruit or seed which aids in dispersal

References

Australian Biological Resources Study. [monograph online]. Flora of Australia: Cumulative glossary for vascular plants. http://155.187.10.12/glossary/glossary.html. Accessed June 20,1996.

Greenlee, J. 1992. *The Encyclopedia of Ornamental Grasses: How to grow and use over 250 beautiful and versatile plants.* New York: Rodale Press. 186p.

Harlow, W.M., E.S. Harrar, and F.M. White. 1979. *Textbook of Dendrology Covering the Important Forest Trees of the United States and Canada.* Sixth edition. New York: McGraw-Hill Book Company. 510p.

Harris, J.G. and M.W. Harris. 1994. *Plant Identification and Terminology: An Illustrated Glossary.* Utah: Spring Lake Publishing. 198p.

Hitchcock, A.S. 1971. *Manual of the Grasses of the United States.* Vol. I and II. New York: Dover Publications. Inc. 1051p.

Jensen, E.C. and D.J. Anderson. 1995. The reproductive ecology of broadleaved trees and shrubs: glossary. Corvallis, OR: Forest Research Laboratory, Oregon State University. Research Contribution 9f. 8p.

Keator, G. 1994. *Complete Garden Guide to the Native Shrubs of California.* San Francisco: Chronicle Books. 314p.

Kruckeberg, A.R. 1982. *Gardening with Native Plants of the Pacific Northwest.* Seattle: University of Washington Press. 252p.

Landis, T.D., R.W. Tinus, S.E. McDonald, and J.P. Barnett. 1990. Containers and growing media. Vol. 2, The Container Tree Nursery Manual. Agric. Handbook. 674. Washington, D.C.: U.S. Department of Agriculture, Forest Service. 88p.

Randall, W.R., R.F. Keniston, D.N. Bever, and E.C. Jensen 1994. *Manual of Oregon Trees and Shrubs.* Corvallis: Oregon State University Book Stores, Inc. 305p.

USDA Forest Service. 1974. *Seeds of Woody Plants in the United States.* Agric. Handb. 450. Washington, D.C.: U.S. Department of Agriculture, Forest Service. 883p.

Acknowledgements

The following illustrations were reprinted from: Gilkey, H.M. and L.J. Dennis. 1980. Handbook of Northwestern Plants. Oregon State University Bookstores, Inc. Corvallis, OR. 507 p. with permission from L.J. Dennis:

Arbutus menziesii (Pacific madrone), *Arctostaphylos nevadensis* (pinemat manzanita), *Ceanothus sanguineus* (redstem ceanothus), *Cornus canadensis* (bunchberry), *Hieracium albifolium* (white hawkweed), *Holodiscus discolor* (oceanspray), *Linnaea borealis* (twinflower), *Lonicera involucrata* (bearberry honeysuckle), *Rhamnus purshiana* (cascara), *Rhododendron macrophyllum* (Pacific rhododendron), *Rosa gymnocarpa* (baldhip rose), *Rosa nutkana* (Nutka rose), *Rubus parvaflorus* (thimbleberry), *Smilacina racemosa* (false Solomon's seal), *Symphoricarpus albus* (snowberry), *Vaccinium ovatum* (evergreen huckleberry), *Vicia americana* (American vetch), *Xerophyllum tenax* (beargrass)

The following illustrations were reprinted from: Randall, W.R., R.F. Keniston, D.N. Bever, and E.C. Jensen. 1994. Manual of Oregon Trees and Shrubs. Oregon State University Book Stores, Inc., Corvallis, OR. 305 p. with permission from E.C. Jensen and OSU Bookstores:

Acer macrophyllum (bigleaf maple), *Arctostaphylos patula* (greenleaf manzanita), *Betula occidentalis* (western water birch), *Chrysolepis (Castanopis) chrysophylla* (golden chinkapin), *Cornus nuttallii* (Pacific dogwood), *Crataegus douglasii* (Douglas hawthorn), *Gaultheria shallon* (salal), *Mahonia aquifolium* (shining Oregon grape), *Mahonia nervosa* (Oregon grape), *Physiocarpus capitatus* (Pacific ninebark), *Physiocarpus malvaceus* (mallow ninebark), *Philadelphus lewisii* (mockorange), *Prunus emarginata* (bitter cherry), *Quercus garryana* (Oregon white oak), *Quercus kelloggii* (California black oak), *Sambucus cerula* (blue elderberry), *Shepherdia canadensis* (buffaloberry), *Vaccinium membranaceum* (big huckleberry), *Vaccinium parvifolium* (red huckleberry)

The following illustrations by Gretchen Bracher were reprinted as follows with permission from Forest Research Laboratory, Oregon State University:

Alnus rubra (red alder) from Jensen, E.C., D.J. Anderson, and D.E. Hibbs. 1995. The reproductive ecology of broadleaved trees and shrubs: red alder, *Alnus rubra* Bong. Forest Research Laboratory, Oregon State Univeristy, Corvallis, OR. Research Publication 9c. 7 p.

Acer circinatum (vine maple) from Jensen, E.C., D.J. Anderson, J.C. Tappeiner, and J.C. Zasada. 1995. The reproductive ecology of broadleaved trees and shrubs: vine maple, *Acer circinatum* Pursh. Forest Research Laboratory, Oregon State Univeristy, Corvallis, OR. Research Publication 9b. 7 p.

Rubus spectabilis (salmonberry) from Jensen, E.C., D.J. Anderson, and D.E. Hibbs. 1995. The reproductive ecology of broadleaved trees and shrubs: salmonberry, *Rubus spectabilis* Pursh. Forest Research Laboratory, Oregon State Univerty, Corvallis, OR. Research Publication 9e. 7 p.

The following illustrations were reprinted from: Hitchcock,C.L. and A. Cronquist. 1973. Flora of the Pacific Northwest: An illustrated manual. University of Washington Press. Seattle, WA. 730 p. with permission from University of Washington Press:

Alnus viridis spp. sinuata (Sitka alder), *Ribes lacustre* (swamp gooseberry), *Viburnum edule* (highbush cranberry)

The following illustrations were reprinted from: Kruckeberg, A.R. 1982. Gardening with Native Plants of the Pacific Northwest. University of Washington Press. Seattle, WA. 252 p. with permission from University of Washington Press:

Amelanchier alnifolia (serviceberry), *Arctostaphylos uva-ursi* (kinnickinnick), *Ceanothus velutinus* (snowbrush), *Fragaria vesca* (woods strawberry), *Pachistima myrsinites* (Oregon boxwood), *Petasites frigidus* (coltsfoot), *Populus trichocarpa* (black cottonwood), *Purshia tridentata* (antelope bitterbrush), *Rhododendron albiflorum* (Cascade azalea), *Sorbus sitchensis* (mountain ash).

The following illustrations were reprinted from: Schmidt, M.G. 1980. Growing California Native Plants. University of California Press. Berkeley, CA. 366 p. with permission from the Regents of the University of California:

Asarum caudatum (wild ginger), *Cornus stolonifera (sericea)* (red-osier dogwood), *Dicentra formosa* (Pacific bleeding heart), *Eriogonum umbellatum* (sulpher buckwheat), *Lupinus latifolius* (broadleaf lupine)

The following illustration was reprinted from: Hermann, F.J. 1970. Manual of the carices of the Rocky Mountains and Colorado Basin. USDA Forest Service. Ag. Handbook No. 374.

Carex utriculata (rostrata) (beaked sedge)

The following illustrations were reprinted from: Sampson, A.W. and B.S. Jespersen. 1981. California Range Brushlands and Browse Plants. University of California Division of Agricultural Sciences. California Agricultural Experiment Station. Extension Service. Berkeley, CA. 162 (Copyright - University of California Board of Regents. Used with permission.):

Artemisia tridentata (big sagebrush), *Cercocarpus ledifolius* (curlyleaf mountain mahogany), *Populus*

tremuloides (quaking aspen), *Salix scouleriana* (Scouler's willow)

The following illustrations were reprinted from: Hitchcock, A.S. 1971. Manual of the Grasses of the United States. Vol. I and II. Dover Publications, Inc. New York. 1051 p. with permission from Dover publications.

Bromus vulgaris (Columbia brome), *Deschampsia atropurpurea* (mountain hairgrass), *Hordeum brachyantherum* (meadow barley), *Koeleria cristata* (Junegrass), *Melica harfordii* (Harford's melic)

The following illustrations were reprinted from: USDA Forest Service. 1988. Range Plant Handbook. Dover Publications, Inc. New York. 816p. with permission from Dover publications.

Acer glabrum (Rocky Mt. maple), *Anaphalis margaritacea* (pearly everlasting), *Aquilegia formosa* (red columbine), *Balsamorhiza sagittata* (arrowleaf balsamroot), *Bromus carinatus* (California brome), *Ceanothus cuneatus* (buckbrush), *Ceanothus prostratus* (squawcarpet), *Cercocarpus montanus* (true mountain mahogany), *Danthonia californica* (California oatgrass), *Danthonia intermedia* (mountain oatgrass), *Deschampsia caespitosa* (tufted hairgrass), *Eriogonum nudum* (barestem buckwheat), *Festuca idahoensis* (Idaho fescue), *Heracleum lanatum* (cow-parsnip), *Potentilla fruticosa* (shrubby cinquefoil), *Ribes cereum* (squaw current), *Spirea douglasii* (hardhack), *Stipa occidentalis* (western needlegrass), *Symphoricarpus oreophilus* (mountain snowberry), *Wyethia amplexicaulis* (mules ear)

The following illustrations were reprinted from: Haeussler, S., D. Coates, and J. Mather. 1990. Autecology of common plants in British Columbia: A literature review. British Columbia Ministry of Forests. FRDA Report-158. 272 p. with permission from B.C. Ministry of Forests:

Alnus incana (thinleaf maple), *Calamagrostis rubescens* (pinegrass),

Epilobium angustifolium (fireweed), *Rubus idaeus* (red raspberry), *Sambucus racemosa* (red elderberry)

The following illustration was reprinted from: USDA. 1971. Common Weeds of the United States. Dover Publications, Inc., New York. 463 p.:

Rhus glabra (smooth sumac)

The following illustrations were reprinted from: Wasser, C.H. 1982. Ecology and Culture of Selected Species Useful in Revegetating Disturbed Lands in the West. USDI, Fish and Wildlife Service, Pub. No. FWS/OBS 82/56. Washington, DC: U.S. Government Printing Office. 347 p.:

Achillea millefolium (yarrow), *Agropyron spicatum* (bluebunch wheat grass), *Mahonia repens* (creeping Oregon grape), *Pinus ponderosa* (ponderosa pine), *Poa sandbergii* (Sandberg's bluegrass), *Prunus virginiana* (chokecherry), *Rosa woodsii* (Wood rose)

The chapter on general propagation techniques was adapted from *Seedling Propagation, Volume 6, The Container Tree Nursery Manual.* (Landis, T.D., Tinus, R.W., McDonald, S.E., and Barnett, J.P. Agric. Handbk. 674., U.S.D.A., Forest Service., manuscript in progress) with generous permission from Tom Landis (Western Nursery Specialist, USDA Forest Service, Cooperative Forestry, P.O. Box 3623, Portland, OR 97208). We gratefully acknowledge this contribution.

Illustrations in the glossary were drawn by Gretchen Bracher, Oregon State University.

Thanks to Susan Buis and Erric Ross for their review of the manuscript and to the staff at OSU Press for their enthusiasm and assistance.

We also wish to acknowledge Fred Zensen of the USDA Forest Service (Region 6) and the Nursery Technology Cooperative at Oregon State University for their support of the initial research for this book.

Index